Dietary Egg

NIPA® GENX ELECTRONIC RESOURCES & SOLUTIONS P. LTD.
New Delhi-110 034

About the Editors

Kiran M, Ph.D. is working as Associate Professor at Veterinary College, Bidar, Karnataka Veterinary Animal and Fisheries Sciences University (KVAFSU) Karnataka. After completing B.V.Sc & A.H from Veterinary College Bengaluru obtained M.V.Sc with Junior Research Fellowship (ICAR) and Ph.D with Senior Research Fellowship (DST) in Livestock Products Technology (meat science) from Sri Venkateswara Veterinary University, Tirupati, India. His research interests include understanding and improvement of meat quality, especially tenderness using biochemical, ultrastructural, and proteome characterization. He has published more than 25 papers in peer-reviewed national and international journals. He has presented his research findings at various national seminars and won several best oral/poster awards and also presented his research finding at international seminars held at Izmir (Turkey), Colombo (Srilanka) and Cork (Ireland). He is actively involved as editorial member and reviewer of many peer reviewed national and international journals. Currently apart from routine teaching and extension activities he is handling project on edible meat packaging sponsored by Dept. Science and Technology.

S. Wilfred Ruban, Ph.D., Wilfred Ruban is currently working at Karnataka Veterinary Animal and Fisheries Sciences University, Karnataka as Associate Professor and Head and has his Ph.D from Madras Veterinary College, Chennai. He has a excellent academic and industrial experience of over 20 years and has contributed significantly to meat science. He is actively involved in teaching undergraduate and postgraduate courses related to Livestock Products Technology. The research activities of Dr. Ruban include food safety and antimicrobial resistance in foods of animal origin. He has published his work in both national and international journals. He is actively involved in mentoring both MVSc and PhD students.

Deepika Jamadar, M.V.Sc, Deepika Jamadar, M.V.Sc completed her post-graduation with specialization in the Livestock Products Technology from the College of Veterinary Sciences, Guru Angad Dev Veterinary and Animal Sciences University, Ludhiana, Punjab. Throughout her academic period, Dr. Jamadar Deepika developed a strong foundation in Livestock Products Technology and gained hands-on experience by attending many workshops.

Dietary Egg
Nutritional and Commercial Facets

Kiran M.
Associate Professor
Veterinary College, Bidar
Karnataka Veterinary Animal and Fisheries Sciences University (KVAFSU)
Karnataka

Wilfred Ruban S.
Associate Professor and Head
Karnataka Veterinary Animal and Fisheries Sciences University
Karnataka

Deepika Jamadar
Specialization in Livestock Products Technology
College of Veterinary Sciences
Guru Angad Dev Veterinary and Animal Sciences University
Ludhiana, Punjab

NIPA® GENX ELECTRONIC RESOURCES & SOLUTIONS P. LTD.
New Delhi-110 034

NIPA® GENX ELECTRONIC RESOURCES & SOLUTIONS P. LTD.

101,103, Vikas Surya Plaza, CU Block
L.S.C.Market, Pitam Pura, New Delhi-110 034
Ph : +91 11 27341616, 27341717, 27341718
E-mail: newindiapublishingagency@gmail.com
www: www.nipabooks.com

For customer assistance, please contact
Phone: + 91-11-27 34 17 17
Fax: + 91-11-27 34 16 16

Print ISBN: 978-93-95763-88-2

ebook ISBN: 978-93-58870-25-1

Composed and Designed by NIPA®.

Preface

This book ***Dietary Egg: Nutritional and Commercial Facets*** provides comprehensive coverage of the Science and Technology of egg for people working in the egg industry along with students and researchers to learn about rapidly booming egg industry. The book covers all areas of egg from initial egg formation to final processing. The goal of this book is to provide thorough information of egg industry. The coverage includes major egg production practices, handling, and nutritional composition of eggs, egg handling storage and preservation. Even though there are many books on this subject but there was long felt necessity to get all information at one place. To meet this demand this book is made concisely to meet such expectation. The best efforts have been made to compile updated information on the topics covered using appropriate tables and figures. This book attempts to highlight the importance of eggs, structure and composition, feeding, management, housing and egg processing.

I would like to thank number of individuals who contributed exhaustive and elaborate write ups on topics during process of writing this book. The development of egg science has been an effort involving many talented individuals. It would be impossible to mention them all here. I thank all of them for contribution to this field.

If you have any comments or suggestions I would appreciate hearing from you.

Editors

List of Contributors

Bhaskar Reddy GV- Assistant Professor and Head, Department of Livestock Products Technology, College of Veterinary Science, Tirupati, Andhra Pradesh

Deepika Jamadar- Assistant Professor, Department of Livestock Products Technology Veterinary College Bidar, Karnataka Veterinary Animal and Fisheries Sciences University Bidar, Karnataka

Kiran M- Associate Professor and Head, Department of Livestock Products Technology Veterinary College Bidar, Karnataka Veterinary Animal and Fisheries Sciences University Bidar, Karnataka

K Sudheer - Assistant Professor, Department of Livestock Products Technology, College of Veterinary Science, Tirupati, Andhra Pradesh

Rajendra Kumar K- Assistant Professor, Department of Poultry Technology, College of Poultry Production and Management, Hosur, Tamilnadu

Sushant Handage- Assistant Professor, Department of Veterinary and Animal Husbandary Extension Eduaction, Veterinary College Gadag, Karnataka Veterinary Animal and Fisheries Sciences University, Bidar, Karnataka

Vidyasagar- Assistant Professor, Department of Livestock Production Management, Veterinary College Bidar, Karnataka Veterinary Animal and Fisheries Sciences University, Bidar, Karnataka

Wilfred Ruban S- Associate Professor and Head, Department of Livestock Products Technology, Veterinary College Bangalore, Karnataka Veterinary Animal and Fisheries Sciences University, Bidar, Karnataka

Contents

1

Overview of Poultry Industry

Kiran M

Abstract

In just 40 years, India's chicken industry has witnessed a dramatic transformation, evolving from a simple interest into a significant commercial agribusiness. Splendid annual growth rates of 4-6% may be attributed to the standardisation of procedures in nutrition, housing, management, and disease control, as well as the development of high producing layer varieties in India. Consistent with the rise in output, the yearly per capita availability of eggs also rose to 90. Almost 95% of India's total egg output comes from chickens, with the remaining 5% coming from ducks and other species. Feed comprises about 65% - 80% of the total cost of producing layers. The cost of compounded feed is often determined by the proportion of corn to protein meal, such as soybean meal. Biosecurity, immunization, and treatment are the three cornerstones of disease control in order to reduce the prevalence of illness in chicken. Despite several obstacles, the Indian poultry industry has shown remarkable development throughout the years. India's poultry industry expects growth and mechanisation to keep up with rising egg demand.

1.1 Introduction

Poultry is one of the fastest-growing areas of India's agricultural industry, with an annual growth rate of around eight percent. Over a span of four decades, the chicken industry in India has seen a paradigm shift in structure and operation, transforming from a simple backyard occupation to a significant commercial agri-based enterprise. Constant efforts in upgrading, modifying, and using new technologies opened the path for the multiplication and diversification of poultry industries. The expansion is not just quantitative, but also qualitative, sophisticated, and productive. This transition has necessitated substantial expenditures and increases in breeding, hatching, raising, and processing. The rise of the poultry industry in India is also characterised by the expansion of chicken farms. The structure of the poultry industry in India differs by area. While independent and generally small-scale farmers account for the majority of production, integrated large-scale producers account for an increasing proportion of output in certain locations.

1.2 Current Scenario

India is the third-largest producer of eggs and chicken meat in the world. In India, 260 million layers generate around 3.4 million tonnes (74 billion) of eggs per year, while 3000 million broilers create about 3.8 million tonnes of chicken meat. The poultry industry employs more than 4 million people directly or indirectly and contributes over 70,000 crores of rupees to the national GDP. Every year, a byproduct called chicken litter, a useful organic fertiliser, is generated in quantities of around 2-2.5 million tonnes. In several regions of the nation, there is a concentration of the poultry sector. Leading the nation are the States of Andhra Pradesh, Telangana, and Tamil Nadu, followed by Maharashtra, Punjab, and West Bengal.

The huge range in production levels is the main cause of the very uneven availability of eggs across the nation. The urban population consumes the majority of the eggs produced, whereas the availability of industrially produced eggs and meat in rural and tribal regions is quite limited. Despite its rapid expansion, the poultry industry has recently experienced many setbacks due to factors such as rising feed prices, the emergence of new or reemerging diseases, fluctuating market prices for eggs and broilers, etc. These issues must be resolved if the poultry industry is to become a sustainable business.

Table 1.1 : Top 10 egg producing states in India (Lakhs)

State/Union Territory	**2004-05**	**2009-10**	**2014-15**	**2019-20**
Andhra Pradesh	158040	193958	130958	219275
Tamil Nadu	63948	108476	159253	200216
Telangana	-	-	106185	148055
West Bengal	28877	36978	48136	97350
Karnataka	17719	29094	43968	66511
Haryana	14816	38453	45790	66153
Maharashtra	34362	38640	50792	63713
Punjab	36800	32828	42642	56388
Uttar Pradesh	9018	10596	20776	34049
Bihar	7894	11002	9845	27408
All India	**452009**	**602671**	**784839**	**1143831**

*Source:*https://www.statista.com/statistics/1357100/india-egg-production-share-by-major-state/

1.3 Production Systems

The fast growth of the poultry industry has been linked to technical advancement and an increase in the size of production facilities. More precisely, the evolution has entailed a shift in focus from traditional small-scale production utilising indigenous breeds that were developed for both meat and egg production to intense commercial production systems employing hybrid birds. The expansion and development of the commercial chicken industry begin with the introduction of enhanced, foreign genetic material. The new strains of birds are often less resilient and resistant to endemic illnesses than native species. Without supplementary inputs like specially formulated concentrate diets, better housing, management, and veterinary care, the higher productive potential cannot be reached. However, the addition of new genetic material serves as the framework for all subsequent technological advancements.

1.4 Feed Resources

The use of high-quality birds, a cosy atmosphere, and the availability of excellent feed—the last of which is the most expensive of all other inputs—are the three main factors that determine the success of poultry production. Feed makes up 65–70% of the cost of producing broilers and 80–85% of the cost of producing layers. The cost of compounded feed is often determined by the cost of maize, which is a common grain that is combined with protein meals like soybean meal. From 9.65 million tonnes in 1989–1990 to barely 24.4 million tonnes in 2015, maize production grew. Similar to this, the output of soybean meal climbed from 3.52 million tonnes in 1999–2000 to 11.35 million tonnes in 2015. The annual average rise in maize supply has been 3.8%, which is significantly less than the growth rate of the production of eggs or beef. Therefore, there is a need to either research the utility of various alternative energy and protein-rich feedstuffs to maize and soybean meal, respectively, in poultry diets, or to boost the production of maize and soybean.

1.5 Disease Management

The control of poultry diseases is crucial to the development of the sector. In commercial farms, birds are raised under open-sided housing and kept in ideal management circumstances. Under the guidance of veterinarians, birds are raised. Regular vaccination is used to safeguard the bird from illnesses. Despite all the precautions, the avian influenza epidemic last year severely harmed India's chicken business. After the consumption of poultry meat and eggs declined for nearly six months, the sector experienced significant trade losses.

1.6 Food Safety

In order to reduce the use of antibiotics in animal feed, herd health concerns must be significantly improved. However, in order to avoid the risks of antibiotic resistance in humans, it is also important to handle antimicrobial medications sensibly and professionally. Antimicrobial medications may be used for effective therapeutic and preventive purposes, but the pulsing or continuous use of antibiotics in feed has been seriously questioned due to effects on the intestinal microbiota and the gastrointestinal barrier harmonic function, in addition to public health concerns. Because of disease resistance and antibiotic residues in the food chain, there is a global concern about using as little antibiotics as possible in chicken. In this situation, relevant alternatives that could be advantageous and cost-effective need to be investigated. Probiotics, gut acidifiers, immunomodulators, symbiotics, organic acids, and other goods of this kind are widely accessible on the market, but they still require more study. The preservation of human health and improvement of quality of life depend heavily on the availability of safe food. Whether domestically produced and eaten, imported, or exported, safe food is crucial. The creation of safe food also gives a chance for revenue generating and market access.

1.7. Marketing

Although commercial production of eggs and chicken meat has been well-standardized according to scientific principles, marketing of eggs and broiler meat is only partially structured in metropolitan areas. Eggs are still being carried in open containers and without refrigeration. In India, eggs are a common item that individuals buy for everyday necessities mostly from the store next door. Egg prices are increased by 10-15% over the real sale price at the producer's location as a result of the two to three stages of distribution that take place through wholesalers, sub-dealers, retailers, etc. Broilers are either butchered or sold live at the point of sale. Sometimes, without any regard for sanitation, the birds are dressed and put on display for sale outside. Similar to that, eggs are sold in the open without regard for quality preservation. The biggest obstacle to price stability is seasonal changes in egg and meat consumption and demand. The fluctuations might occasionally reach up to 25% to 30% in just 3 to 4 weeks.

1.8. Processing and Export

Some states have established state-of-the-art egg processing factories, with daily turnover capacities of 0.7 million to 0.8 million eggs. The production of egg powder, lysozyme, egg weight powder, and other products adheres to strict guidelines. India's egg powder has found widespread success in the European

Union, Japan, and the Far East. However, many more egg processing units are required to access the global market. India's convenient location has led to claims that the country may serve as a distribution hub for shell eggs destined for the Middle East and far eastern regions. This means that there is a lot of room for growth in India's export of shell eggs to these nations.

1.9. Organic Production

Due to the widespread use of chemical pesticides and fertilisers in modern agricultural practises, there has been a recent uptick in interest in organic food items amongst consumers. Integrated, humane, ecologically and economically sustainable agricultural production methods that provide adequate amounts of crop, animal, and human nourishment, pest and disease protection, and a suitable return on human and other resources utilised; this is the goal of organic farming. Renewable resources sourced from nearby areas or farms are prioritised, as are systems that allow these resources to regulate themselves, such as ecological and biological ones. All available measures are taken to lessen the plant's reliance on chemical and other external inputs. Due to its emphasis on ecosystem management rather than the use of synthetic fertilisers and pesticides, organic farming is also known as ecological farming. In India, free-range farming meets the criteria for organic status if the birds are not given any antibiotics or other antibiotics in their feed.

1.10. Challenges and Way Ahead

A number of challenges have emerged for the poultry industry in recent years, including the rising cost of feed, the appearance of new or re-emerging diseases, the fluctuating market price of egg and broilers, etc., all of which must be overcome in order to establish the poultry sector as a sustainable business. Concerns over animal cruelty and pollution from chicken farms have grown in recent years. Despite a number of setbacks over the years, India's poultry industry has continued to show remarkable development. Poultry production in India is expecting growth and industrialisation in response to rising demand for chicken egg and meat. The nutritional and economic situation of rural people can be improved if home chicken raising is adopted. Future obstacles will not be a problem because of the growing body of knowledge and breakthroughs in several areas of poultry, therefore the industry here has high hopes for the future.

2

Egg Formation

Kiran M

Abstract

The oviduct of the chicken is a special organ where an ovulated yolk develops into a whole egg. During egg development and oviposition, ovarian hormones cause cellular and biochemical changes in the oviducts. Estradiol controls ovulation, the formation of oviducts, the accumulation of yolk in follicles, and folliculogenesis. Additionally, estradiol stimulates the production of the genes for the proteins found in egg whites and glandular growth. The ovary's yolk releases during ovulation, and progesterone also stimulates the growth of the oviductal glands.

2.1 Introduction

The biochemical environment for egg production and ovulation egg fertilisation is provided by the poultry oviduct. The chickens have two ovaries and an oviduct when they are hatched, but the growth of the right ovary and oviduct stops and regresses over time. The left ovary and oviduct are still functioning and help produce eggs. The oviduct is a long tubular structure that is divided into five functionally and histomorphologically distinct segments: the infundibulum, which is the site of fertilisation, the magnum, which produces egg white components, the isthmus, which forms the egg shell membranes, the shell gland or uterus, which produces the calcified eggshell, and the vagina (oviposition or egg laying). The components of the egg are released and deposited from certain sections of the oviduct as the ovum travels the length of the oviduct after ovulation. A full egg is created in around 24-28 hours after the yolk enters the oviduct. Each section of the oviduct that the egg passes through either creates a part of the egg or performs an essential non-secretory function. Oviductal functions, in addition to environmental, nutritional, and pathological factors, control egg production and quality. The development of the egg inside the oviduct is a highly complicated process that is governed by hormones and genetics. The biological processes and genes involved in egg production are many. This chapter's objective is to present up-to-date knowledge on the function of hormones, genes, and proteins, as well as how they combine to cause histomorphological and biochemical changes in the oviduct segments necessary for egg production.

2.2 Histomorphology and Functions of the Oviduct

In hens, there are two types of infundibulums: membranous and muscular. Both envelop the whole ovary. The muscular infundibulum is coated with ciliated cells and serves as a pathway for the yolk inside the oviduct, while the membrane infundibulum covers the ovarian cluster. The egg stays in the infundibulum for about 15 to 30 minutes before moving down into the magnum, where albumen is laid down all around it. Therefore, any possible ovum fertilisation also takes place at the infundibulum. The egg-white proteins that encircle the yolk are produced by the magnum, the biggest section of the oviduct. While the ciliated epithelial cells assist in egg transport, the glandular epithelial cells of the magnum synthesis the various egg-white proteins, store them, and only release them for 2-3 hours when the egg is present in it. The egg white, which is high in protein, serves as the embryo's primary food supply while it is developing. Additionally, it contains a few antibacterial proteins that guard the embryo against harmful microorganisms. The weight of the egg and the hatchling are determined by the albumen, which makes up more than 60% of the whole egg. The egg descends later and spends the next one to two hours in the isthmus, which connects the shell gland and the magnum. The egg albumen is surrounded by the outer and inner eggshell membranes (ESM) in the isthmus. The fibrous networks that make up the eggshell membranes maintain the jelly egg white in place and serve as the site where eggshell mineralization first begins. The egg travels within the shell gland after being encased by the ESM, where it remains for around 18 to 22 hours, during which time calcite crystals are deposited to create the eggshell on the ESM. Since the eggshell contains 95% calcium by composition, it serves as the developing embryo's principal calcium supply. The structure of the eggshell prevents external microorganisms from entering the egg while enabling air to circulate around the egg to allow the inchoate embryo to breathe. The egg is eventually briefly kept in the vagina when the eggshell has fully mineralized. Some birds finish the pigmentation of their eggs in the vagina before laying them.

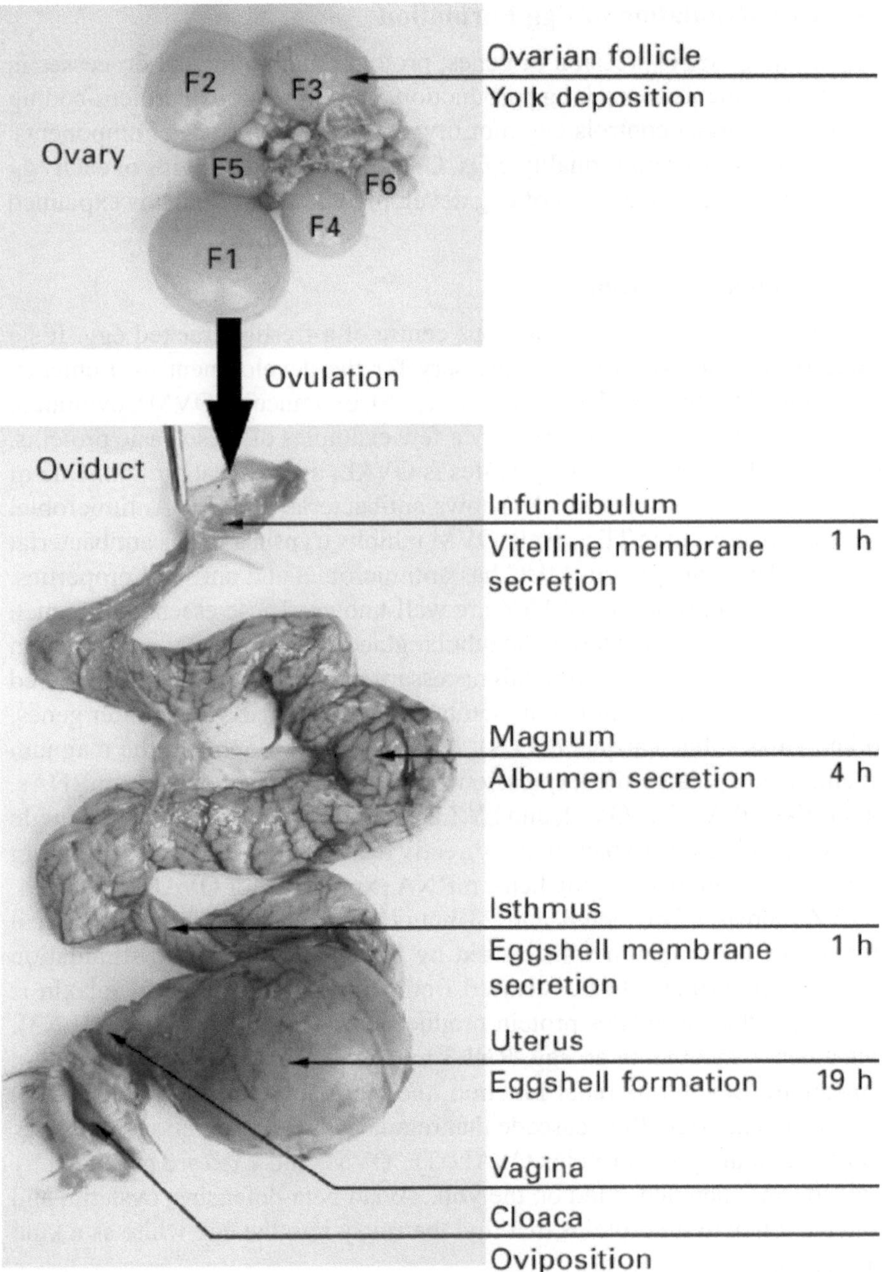

Figure 2.1: Formation of Eggs in Reproductive tract of Hens

Source: Nys, Y., & Guyot, N. (2011). Egg formation and chemistry. In Improving the safety and quality of eggs and egg products (pp. 83-132). Woodhead Publishing Ltd.

2.3 Genetic Regulation of Egg Formation

The spatiotemporal expression of genes, proteins, and biological processes in the oviduct segments controls egg production. The expression of protein-coding genes in the oviduct controls egg motility, the deposition of egg components, and the production of high-quality eggs. On the basis of the origins of each egg component, the genetic control of egg development in the oviduct is explained below.

2.3.1 Albumen Formation

Albumen is the protein-rich gelatinous centre of a freshly cracked egg. It's a mixture of almost 148 proteins necessary for the development of a chicken embryo. Ovalbumin (OVAL), conalbumin (TF), ovomucoid (OVM), ovomucin (MUC), and lysozyme (LYZ) are only a few examples of these basic proteins. About 54% of the protein in egg whites is OVAL, a structural protein. One of OVAL's homologs, ovalbumin X, shows antibacterial activity. Antimicrobial properties are present in TF as well. OVM inhibits trypsin and has antibacterial properties. The mucoprotein MUC has antimicrobial and antiviral properties. The antibacterial properties of LYZ are well-known. These essential albumen proteins are mostly generated in the tubular gland cells of the magnum. Protein precursors, including the amino acids necessary for their synthesis, are delivered to gland cells across the epithelial membrane by specialised transporter genes, also known as solute carriers (SLCs). During egg development, the magnum epithelium upregulates the expression of a large number of SLC mRNAs. Proteins like OVAL, TF, OVM, and LYZ are continually synthesised in a single cell type (gland cells) at a pace that is directly related to their numbers in the egg white. In the magnum of laying hens, mRNA expression for OVAL, TF, OVM, and LYZ increases between 4 and 23 hours post-ovulation. The mechanical distention of the magnum wall caused by the egg's entry is the stimulation that causes the proteins to be released from storage. The chemical relaxin is one example that stimulates protein production from epithelial cells (RLN3). When a hen is incubating an egg, RNL3 mRNA is more heavily expressed in the magnum. Beyond its renal function, the renin-angiotensin system (RAS) is involved in the signalling cascade that regulates protein secretion. Secretory granules containing the proteins OVAL, TF, OVM, and LYZ are produced by the glands and then deposited on the yolk. Avian beta-defensins, cystatin, and avidin are a few more proteins that find their way into the egg white as a kind of protection.

2.3.2 Eggshell Membrane

The membranes of eggshells are fibrous networks structured in outer and interior layers and interlinked with fibres to form a densely interwoven fibrous meshwork. This meshwork offers the nucleation locations for eggshell mineralization to begin. A disruption in the production and structure of these cross-linked fibres can have a deleterious effect on the strength of the eggshell. When the egg is in the isthmus, the expression of numerous genes and proteins is crucial to the creation of the ESM. The primary fibrous components of the ESM are collagens. Collagen X (COL10A1) mRNA expression is elevated in the isthmus of laying chickens. The collagen X proteins are homotrimers of -1 chains that are released by the tubular gland cells of the isthmus and are responsible for maintaining the structural integrity of the ESM. In addition to collagens, fibrillin-1, cysteine-rich eggshell membrane protein (CREMP), lysyl oxidases, quiescin Q6 sulhydryl oxidase 1 (QSOX1), and thioredoxin are required for ESM development. Only within the isthmus is the microfibrillar glycoprotein fibrillin-1 overexpressed. Fibrillin-1 provides the ESM its elasticity. The majority of cysteine in the ESM is derived from CREMPs, which are highly expressed in the isthmus. The CREMP has antimicrobial properties in the egg. Lysyl oxidases are enzymes present in the ESM that facilitate the production of cross-links between collagen and ESM fibrillar proteins. The QSOX1 protein also promotes the formation of ESM meshwork and controls the ESMs' structural integrity. Between fibrillar proteins, thioredoxin catalyses the creation of disulfide cross-links.

2.3.3 Eggshell Biomineralization

Eggshell membranes are strongly cross-linked fibrous networks composed of outer and interior layers of interwoven fibres. The mineralization of an eggshell begins at the nucleation sites provided by this meshwork. The strength of an eggshell is affected when the production and structure of its cross-linked fibres are disturbed. Essential to the development of the ESM is the expression of a number of genes and proteins throughout the egg's time in the isthmus. The ESM's primary fibrous components, collagens, have a crucial structural role. In laying hens, collagen X (COL10A1) mRNA is most abundant in the isthmus. Elastin and scleroprotein (ESM) structure is maintained by collagen X proteins, which are a homotrimer of -1 chains released by tubular gland cells in the isthmus. Collagens aren't the only proteins involved in ESM development; others, such fibrillin-1, cysteine-rich eggshell membrane protein (CREMP), lysyl oxidases, quiescin Q6 sulfhydryl oxidase 1 (QSOX1), and thioredoxin, are essential as well. Microfibrillar glycoprotein fibrillin-1 has isthmic mRNA overexpression. Fibrillin-1 is responsible for the ESM's elastic properties. Most of the cysteine in the ESM is produced by cysteine-rich eukaryotic mRNAs for muscle (CREMPs), which are highly expressed in the isthmus. There is some evidence that the CREMP in the egg has antibacterial properties as well. Cross-links between collagen and ESM fibrillar proteins are mediated by lysyl oxidases, which are enzymes present in the ESM. As a mediator of ESM origin, QSOX1 also controls ESM integrity. Disulfide cross-links between fibrillar proteins are formed thanks to the enzyme thioredoxin.

2.3.4 Ubiquitous Proteins and Egg Formation

Numerous proteases known as matrix metalloproteases (MMPs) are known to break down various extracellular matrix proteins (ECM). The ECM that surrounds cells in the body is degraded and remodelled by MMPs to control cellular growth, proliferation, and differentiation. MMPs are found throughout the oviduct, primarily in the uterus and magnum. Due to the high level of secretion in the uterus and magnum, epithelial growth is necessary. The oviductal epithelium's surrounding ECM is broken down by the MMPs, which also aid in cellular migration, proliferation, and differentiation. Throughout moulting, certain MMPs (MMP-2, -7, and -9) are actively expressed in the oviduct; however, during the transition from immature to adult hens, these MMPs are downregulated. Compared to non-laying and moulting chickens, laying hens express the greatest levels of MMP-1 and -10. Collagen interstitial is broken down by MMP-1 (type I, II, and III). MMP-2 causes angiogenesis by breaking down type IV collagen. Matrilysin, another name for MMP-7, is the enzyme that breaks down proteoglycans, casein, fibronectin, and elastin. The gelatinase MMP-9 stimulates the growth of new vasculatures. A stromelysin enzyme called MMP-10 may break down fibronectins and proteoglycans. The aforementioned MMPs' varied matrix-degrading actions finally guarantee appropriate oviduct reproductive capabilities.

2.4 Hormonal Regulation of Egg Formation

The laying hen's process of producing eggs is a complex one that involves the interaction of several chemicals and hormones. The development of the reproductive system, ovulation, albumen synthesis, eggshell creation, and eventually the oviposition of eggs are all processes in which hormones play a crucial role.

2.4.1 Gonadotropin-releasing Hormone (GnRH)

In hens, the hypothalamic/portal system secretes GnRH in response to photo-stimulation and an increase in the concentration of progesterone. This happens in tandem with the photo-stimulation. Chicken GnRH-I (also written as cGnRH-I) and chicken GnRH-II are the two different chemical variants of the GnRH that are found in avian species (cGnRH-II). Different functions are performed by each of these two versions of the GnRH in bird species. GnRH-I is necessary for stimulating the synthesis and release of anterior pituitary hormones, whereas GnRH-II is engaged in mating and courting behaviour. GnRH-I is essential for stimulating the synthesis and release of anterior pituitary hormones. Catecholamines, vasotocin, vasoactive intestinal peptide, neuropeptide Y, and opioid peptides all have a role in the regulation of GnRH in hens. It was just recently discovered that the GnRH receptor is present in the oviduct of laying hens; nevertheless, its actual purpose in the process of egg creation is still a total mystery.

2.4.2 Gonadotropins

In response to the hypothalamic release of gonadotropin-releasing hormone (GnRH), the anterior pituitary gland produces the gonadotropins known as follicle-stimulating hormone (FSH) and luteinizing hormone (LH). Within the hen, FSH is the hormone that is accountable for the process of granulosa cell growth and recruitment within the tiny follicles. The granulosa layer of the little yellow follicles and the sixth (F6) to third (F3) biggest follicles are the primary targets of the FSH hormone's action. Additionally, it promotes the synthesis of progesterone in the granulosa cells of follicles ranging from F6 to F3. The sustained plasma concentration of FSH is present throughout the entirety of the ovulatory cycle, with the exception of a minute spike that occurs around 12 hours before to ovulation. In contrast to other mammalian species, the LH in hens does not luteinize the follicles; rather, the follicles are engaged in the process of ovulation and the production of steroids. The plasma concentration of the LH reaches its highest point between the hours of 4 and 6 hours before ovulation (which coincides with the peak rise in progesterone), and it is around 11 hours before ovulation that the plasma concentration of the LH is at its lowest point. The growth of preovulatory follicles that are much bigger is the LH's principal objective.

2.4.3 Estrogens

The theca cells found in the tiny follicles are primarily responsible for the production of estrogens. The plasma concentration of estradiol is at its maximum between the hours of 4 and 6 hours before ovulation, however a little increase in oestrogen levels can also be seen between the hours of 18 and 23 hours before ovulation. Estrogen is an essential component in the development of the egg yolk. It does this by increasing the production of vitellogenin and very low-density lipoprotein in the avian liver. These two substances are the principal sources of yolk protein and fat, respectively. Estradiol has the additional impact of making the hypothalamus more sensitive to the positive feedback effect of progesterone. In addition to playing a crucial part in the growth and development of the oviduct, estradiol is also responsible for regulating calcium metabolism, which is necessary for the creation of eggshells and the maturation of secondary sexual characteristics. Ovalbumin, conalbumin, ovomucoid, and lysozyme are the primary components of albumen, which is mostly produced in the tubular gland cells of the magnum and contains the majority of the albumen. Since oestrogen was shown to be involved in the production of these molecules, it is clear that it plays a significant part in the creation of egg white.

2.4.4 Progesterone

The hormone progesterone and the receptor for it are responsible for regulating a woman's fertility. Granulosa cells of the bigger follicles are primarily responsible for the production of progesterone (F1–F3). It takes between four and six hours before ovulation for the plasma concentration of progesterone to reach its highest point. Progesterone is only secreted by the biggest preovulatory follicles while the preovulatory LH surge is occurring because of this fact. This rise in progesterone triggers a positive feedback response in the hypothalamus, which in turn stimulates an increase in the secretion of gonadotropin-releasing hormone (GnRH) into the hypothalamic-pituitary portal system, leading to an increase in luteinizing hormone (LH) production by the anterior pituitary. This LH stimulates the mature follicles to burst, which results in the release of the egg yolk (ovum) (F1). In addition, progesterone is connected to the development of avidin, the constriction of the myometrium, and the construction of eggshells.

2.4.5 Androgen

Both small and big follicles have cells that make androgen in the theca and granulosa layers. Testosterone levels peak 6-10 hours before ovulation, whereas 5-dihydrotestosterone peaks at the same time. The function of androgen in ovulation is still not well understood. Both ovomucoid and ovalbumin gene expressions in the chicken oviduct are discovered to be regulated by androgen. Androgens aid in the maturation of a hen's secondary sexual characteristics, such the size and pigmentation of her comb and wattle.

3

Structure of An Egg

Kiran M

Abstract

Long acknowledged as a superior source of nutrition for humans, bird species' eggs. Innovative studies that have recently revealed the variety of egg components' structures and functions have spurred rising demand for this bioresource to be used to its full potential. These applications are being created to benefit from the nutritional and practical benefits of eggs in food products, as well as their bioactive components, which have the potential to be used as nutraceutical and functional food ingredients with the aim of lowering disease risk and improving human health.

3.1 Introduction

Eggs are composed of three main parts; the eggshell with the eggshell membrane, the albumen or white, and the yolk. The yolk is surrounded by albumen, which in turn is enveloped by eggshell membranes and finally a hard eggshell.

3.2 Eggshell

The eggshell is composed of a foamy layer of cuticle, a calcite or calcium carbonate layer, and two shell membranes. The ultrastructurally the eggshell includes shell membranes, mammillary zone, calcium reserve assembly, palisades, and cuticle, with 7,000 – 17,000 funnel shaped pore canals distributed unevenly on the shell surface for exchange for water and gases

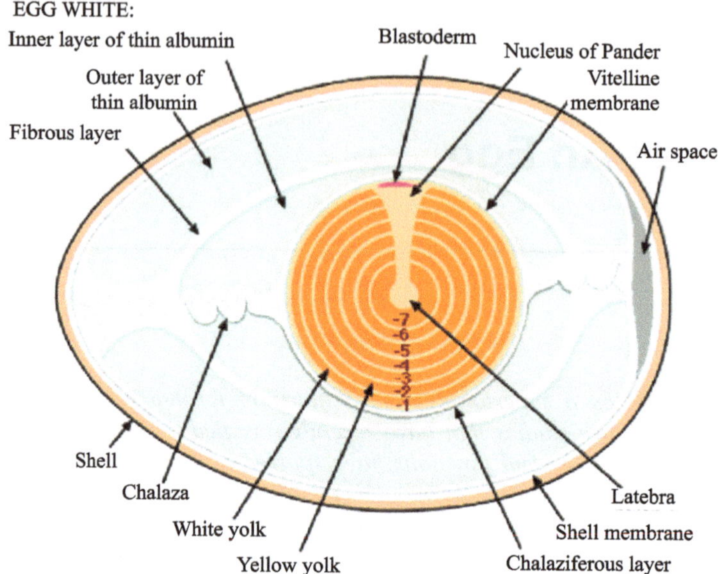

Figure 3.1: Structure of Egg

Source: Adegbenjo, A. O., Liu, L., & Ngadi, M. O. (2020). Non-destructive assessment of chicken egg fertility. Sensors, 20(19), 5546.

3.2.1 Cuticle

The water - insoluble cuticle is the most external layer of eggs; it is about 10 – 30 μm thick and covers the pore canals. It helps protect the egg from moisture and microbial invasion. The cuticle is composed of an inner mineralized layer and an outer layer consisting of only organic matrix. The layer adjacent to the shell has a foamy appearance, whereas the outer layer is more compact. The inner cuticular layer, referred to as the vesicular cuticle, is composed of a matrix - like material containing vesicles of various sizes that contain distinctly granular material with an electron - lucent core and an electron - dense mantle. The outer layer of cuticle is much more compact and homogeneous, does not appear to contain any matrix vesicles, and is referred to as the nonvesicular cuticle.

3.2.2 Shell Matrix

The eggshell matrix consists of the vertical crystal layer, palisade layer (sponge layer), and mammillary knob layer. Calcium carbonate is the major component of inorganic substances. This calcite (the most stable form of calcium carbonate) forms elongated structures termed columns, palisades, or crystallites.

3.2.3 Shell Membrane

The eggshell membrane structure is composed of inner and outer membranes that reside between the albumen and the inner surface of the shell. The inner membrane, also referred to as the egg membrane , has three layers of fibers that are parallel to the shell and at right angles to each other. On the other hand, the outer membrane, also referred to as the shell membrane, has six layers of fiber oriented alternately in different directions.

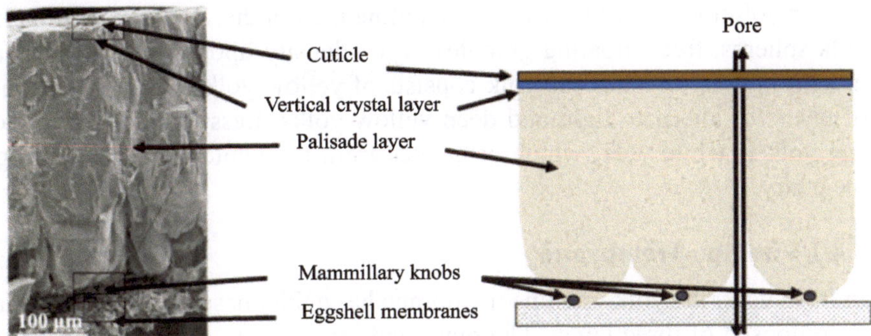

Figure 3.2: Structure of Egg shell

Source: Gautron, J., Stapane, L., Le Roy, N., Nys, Y., Rodriguez-Navarro, A. B., & Hincke, M. T. (2021). Avian eggshell biomineralization: an update on its structure, mineralogy and protein tool kit. BMC Molecular and Cell Biology, 22(1), 1-17.

3.3 Albumen

Albumen or egg white is composed of four distinct layers: an outer thin white next to the shell membrane, a viscous or outer thick white layer, an inner thin white, and a chalaziferous or inner thick layer. The contents of each layer are about 23.3%, 57.3%, 16.8%, and 2.7%, respectively. The proportions may vary, however, depending on breed of the hen, environmental conditions, size of the egg, and rate of production.

3.3.1 Thick and Thin Albumen

In fresh eggs, thick albumen covers the inner thin albumen and the chalaziferous layer, keeping the egg yolk in the center of the egg. The viscosity of thick albumen is much higher than that of thin albumen because of its high content of ovomucin.

3.3.2 Chalaziferous Layers and Chalazae

The chalaziferous layer is a gelatinous layer that directly covers the entire egg yolk. In the long axis of the egg, the chalaziferous layer is twisted at both sides of the yolk membrane, forming a thick rope - like structure termed the chalazae cord. This cord is twisted clockwise at the sharp end of the egg and counterclockwise at the opposite end. The chalazae cord stretches into the thick albumen layer to both sides; thus the egg yolk is suspended in the center of egg.

3.4 Yolk

Yolk is a complex system containing a variety of particles suspended in a protein solution and encircled by a vitelline membrane. The particles include yolk spheres, free - floating granules, low - density lipoprotein globules, and myelin figures. Most of the yolk consists of yellow yolk, which is composed of layers of alternate light and deep yellow yolks. Less than 2% of the total egg yolk is white yolk, which originates from the white follicle maturing in the ovary.

3.4.1 Vitelline Membrane

In fresh yolk, the yolk vitelline membrane has a thickness of about 10 μ m and is composed of three layers The outer and inner layers are each about 5 μ m thick, but the former has a coarse structure with fi ber layers, while the latter has a closer structure with dense net - like fi bers. Diameters of the fi bers are 200 – 600 and 15 nm in the outer and inner layers, respectively.

3.4.2 Yellow Yolk

Yellow yolk consists of two types of lipoprotein emulsion: the deep yellow yolk and the light yellow yolk. The deep yellow yolk is formed in the daytime, and the light yellow yolk is formed at night, leading to the alternate and circular appearance of these yolk layers.

4

Chemical Composition of Eggs

Kiran M

Abstract

The egg is made up of three primary components: the egg white, the egg yolk, and the egg shell. The calcite crystals of the shell are enmeshed in a protein and carbohydrate complex matrix. About 58% of the weight of an egg is made up of the viscous, colourless liquid termed egg white inside the shell. A whole egg is made up of 28–29% yolk, 60–63% albumen, and 9–11% eggshell. Egg white and yolk are the main sources of proteins, but the yolk is virtually entirely the source of lipids. The main element of the eggshell is a mineral.

4.1 Introduction

The egg is made up of three primary components: the egg white, the egg yolk, and the egg shell. The calcite crystals of the shell are enmeshed in a protein and carbohydrate complex matrix. About 58% of the weight of an egg is made up of the viscous, colourless liquid termed egg white inside the shell. About 75% of the weight of an egg's edible component is made up of water, although proteins and lipids make up the majority of its nutritional worth (Table 4.1). There are also trace quantities of minerals, often known as ash, and carbohydrates, which take the form of simple sugars including glucose, sucrose, fructose, lactose, maltose, and galactose.

4.2 Egg Shell and Membrane

Sometimes referred to as a natural composite bioceramic, avian eggshell is composed mostly of inorganic (95%) and organic (3.5%) components. It contains calcium carbonate 95% (w/w), 1.0% (w/w) matrix proteins, and 0.5% (w/w) other materials. The insoluble eggshell layers have a fatty acid concentration between 2 and 4%, which includes fatty acids that are both lipid and lipoprotein bound. With the exception of behenic acid, which is present in considerably higher concentration in the cuticle layer, there is often minimal change in the fatty acid content between the egghell layers. Additionally predominate in the cuticle layer are porphyrin pigments. An essential pigment in chicken and other bird species' eggshells is biliverdin. The average amount of uronic acid, a carbohydrate component of glycosaminoglycans, in an eggshell is 0.024% of its dry weight. About 48% of the hyaluronic acid and 52% of the galactosaminoglycans found in eggshells are glycosaminoglycans. Nearly 90% of the cuticle's solid material is made up of protein (primarily in the form of insoluble proteins and glycoproteins), 3% of which is ash, 5% of which is carbohydrates. A protein and polysaccharide matrix rich in sulfated molecules makes up around 2% of the eggshell's organic matrix, which is made up of about 97% calcium carbonate in the form of calcite crystals. The inner and outer shell membranes are made up of 90% protein, 2% glucose, and ash (dry weight basis). Early research claimed that keratin made up the eggshell membranes, while more current reports claim that type X collagen makes up the majority of the membrane fibres and that this inhibits mineralization. Keratin sulphate, a glycosaminoglycan polyanionic substance believed to be a component of the glycosylated group of type X collagen, is also found in the core of the shell membrane fibres.

4.3. Albumen

The percentage of water in egg albumen varies from 84% to 89% from the outermost to the innermost layers. Proteins make for around 10–11% of the albumen solids, whereas carbohydrates (mainly glucose), lipids, and minerals are minor constituents. Specific egg protein components are related with a variety of biological qualities, such as antibacterial activity, protease inhibitory action, immunomodulatory, anticancer, and antihypertensive actions, vitamin-binding capabilities, and antigenic or immunologic characteristics. Ovalbumin is the predominant protein in albumen, accounting for approximately 54% of total egg white protein. It is a member of the serpin superfamily, which consists of more than 300 similar proteins with various activities, including the primary serine protein inhibitors in human plasma. Ovotransferrin, also known as conalbumin, is comparable to animal serum transfer. Transferrins are a class of bilobal glycoproteins that bind ferric iron with high affinity. Ovotransferrin's formation of an iron complex hinders the development of iron-dependent bacteria. The ovomucoid molecule consists of three unique domains that are linked only by intradomain disulfi de bonds. It is believed that the allergenic potential of chicken egg white ovomucoid is connected to its resistance to heat and digestion, necessitating the development of testing methods even after denaturation. Ovomucin is a sulfated glycoprotein that gives egg albumen its gelatinous form. Insoluble ovomucin dominates the gel-like insoluble portion of the thick albumen of egg white, while soluble ovomucin dominates the outer and inner albumen. Although the biological role of ovoglobulin is unknown, it is believed to be essential for the egg white's foaming feature. The name ovoglobulin refers to ovoglobulin G2 and G3, which make up around 4% of egg albumen proteins, respectively. Ovoglobulin G2 and G3 share several characteristics, such as their molecular weight. Egg white lysozyme consists of 129 amino acid residues and has a near to 10.7-degree isoelectric point. Four disulfide bonds hold together its tertiary structure. Due to its fundamental properties, lysozyme binds to ovomucin, transferrin, and ovalbumin. Two to three times more lysozyme is bound to the chalaziferous layer and chalaza compared to other egg white proteins. Ovomacroglobulin suppresses hemagglutination, exhibits anticollagenase activity, and inhibits various serine, thiol, and metal proteolytic enzymes. The riboflavin-binding protein from chicken egg white is the progenitor of a family that includes other riboflavin- and folate-binding proteins. The high degree of disulfide bridge crosslinking and, in the case of avian proteins, the existence of sections of highly phosphorylated polypeptide chain are peculiar features of these molecules. Ovoinhibitor, the other Kazal inhibitor in egg albumen, is a glycoprotein with six a-type domains and one b-type domain, making it a significantly bigger inhibitor. Basic tetrameric glycoprotein known as avidin is well recognised for its capacity to bind biotin. Albumen includes several enzymes, including those with phosphatase, catalase, and glycosidase activity, in addition to lysozyme.

Table 4.1. General Composition of Egg

Nutrient	**Albumen**	**Yolk**
Protein	9.7 – 10.6	15.7 – 16.6
Lipid	0.03	31.8 – 35.5
Carbohydrate	0.4 – 0.9	0.2 – 1.0
Water	84.3 – 88.8	48
Element		
Sulphur	0.195	0.016
Potassium	0.145 – 0.167	0.112 – 0.360
Sodium	0.161 – 0.169	0.070 – 0.093
Phosphorus	0.018	0.543 – 0.980
Calcium	0.008 – 0.02	0.121 – 0.262
Magnesium	0.009	0.032 – 0.128
Iron	0.0009	0.0053 – 0.011

Source: Réhault-Godbert, S., Guyot, N., & Nys, Y. (2019). The Golden Egg: Nutritional Value, Bioactivities, and Emerging Benefits for Human Health. Nutrients, 11: 684.

Albumen contains very low (0.03% w/w) lipid content. The main fatty acids in albumen lipids are palmitic, oleic, linoleic, arachidonic, and stearic acids. Glucose is the main free sugar, constituting a ~ 0.8 – 1% by weight of albumen. It is usually removed by fermentation prior to drying of egg white to prevent browning caused by Maillard reaction. The major minerals in egg white are sulfur, potassium, sodium, and chlorine; phosphorus, calcium, and magnesium are found in lower quantities, as are various other trace minerals. Albumen does not contain fat - soluble vitamins, but does contain significant proportions of the water - soluble vitamins in egg, including biotin, niacin, and riboflavin.

Table 4.2. Fat Profile of Egg Yolk

Lipid Composition	**Percentage**
Triglycerides	63
Phospholipids	31
Cholesterol	4
Other (including fat soluble vitamins)	2
Fatty Acid Composition	**Percentage**
Saturated fatty acids	35 – 45
Unsaturated fatty acids	55 – 65
Monounsaturated fatty acids	35 – 50
Linoleic acid	10 – 20
Polyunsaturated fatty acids	3 – 5

Source: Réhault-Godbert, S., Guyot, N., & Nys, Y. (2019). The Golden Egg: Nutritional Value, Bioactivities, and Emerging Benefits for Human Health. Nutrients, 11: 684.

4.4. Egg Yolk

Egg yolk comprises 50% solids; lipids (65 – 70% dry basis) and proteins (30% dry basis) are the principal elements of the solid matter. The composition of the solid matter in the vitelline membrane differs from the composition of the yolk itself; it contains more protein (87%) and carbohydrates (10%) than lipids (3%). By centrifugation, egg yolk may be separated into plasma (supernatant) and granule fractions (precipitate). The plasma comprises 78% of the whole liquid yolk; it contains 51% solids composed mostly of lipid (80%), 2% ash, and 18% nonlipid material composed primarily of protein. Livetins and lipoprotein particles, including high - density lipoproteins (HDL), low - density lipoproteins (LDL), and very low - density lipoproteins, account for 16% of egg yolk's protein content (VLDL). The egg yolk solids consist primarily of lipids (32 to 36%). About 65 percent of yolk lipid is triglyceride, 28 to 30 percent phospholipid, and 4 to 5 percent cholesterol. The carbohydrate content of egg yolk ranges between 0.7% to 1.0%, with an estimated 0.3% of free glucose and the remaining linked to glycoproteins and glycolipids. The mineral content of egg yolk is around 1%. Phosphorus is the most abundant mineral in egg yolk, comprising 61% of phospholipid.

5

Egg Production Practices Housing

Vidyasagar

Abstract

Poultry production in India has taken a quantum leap in the last four decades, emerging from use of unscientific farming practices to commercial production systems with state-of-the-art technological interventions. Today, the egg production system became highly specialised and more commercial in production practices. Changes in production practices have generated many technologies in basic disciplines like genetics, nutrition, management and physiology of birds. It's an area to which breeders have been giving increased attention to improve quality of the freshly laid egg.

5.1 Introduction

The egg production in the country has increased from 78.48 billion in 2014-15 to 114.38 billion in 2019-20. India ranks 3rd in egg production in the world. Annual growth rate of egg production was 4.99 per cent during 2014-15, there after there has been a significant improvement in the egg production with 10.19 per cent growth registered in 2019-20 over the previous year. The per capita availability of egg was 86 per annum in 2019-20. Less than a generation ago the main supply of eggs was from small backyard unit. Processors, handlers, and distributors of eggs and egg products need an understanding of egg-production practices to cope more intelligently with problems of product quality and character. In same way, food scientist involved with quality control, new product development and composition studies can benefit from the knowledge how egg is produced. Therefore the purpose of this chapter is to briefly sketch about egg- production practices.

5.2 Housing Practices

5.2.1 Free Range Eggs

This system is normally practiced under village areas which is also known as backyard system. Here, eggs produced by hens that have access to outdoors in accordance with weather, environmental or state laws. In addition to consuming a diet of grains, these hens may forage for wild plants and insects and are sometimes called pasture-fed hens. They are provided floor space, nesting space and perches. Its a oldest system and adopted only when adequate land is available. Stocking density under this system is 300-400 birds per hectare.

5.2.2 Semi Intensive System

Eggs laid by hens living half way reared in houses and half way on ground or range. Here, Birds are confined to houses in night or as per the need, they are also given access to runs. Houses may be simple house, thatched roof, littered earth floor or slatted which provides protection against inclement weather predators and shade. Stocking density under this system is 4-5 birds m.sq. in houses.

5.2.3 Intensive System

It is estimated that more than 60 percent of the world's eggs are produced by this system of rearing. Here the birds are reared in suitable housing system and do not have access to outdoor. Birds are provided with balanced feed and water and scientific management is possible.

5.2.3.1 Deep Litter System

Poultry birds are kept in large pens on floor with Easy access for feed, water, egg collection. Floor is covered with litters, such as straw, saw dust or leaves up to depth of 2-3 inches. Layers are provided with 2 Sq.ft floor space per bird.

Advantages

1. Birds and eggs are safety as enclosed in deep litter intensive pen, which has strong wire netting or expanded metal.
2. Built-up deep litter also supplies some of the food requirements of the birds. They obtain "Animal Protein Factor" from deep litter.
3. The level of worm infestation is much lower with poultry kept on good deep litter than with birds (or chicken) in bare yards.
4. It is a valuable insulating agent, the litter maintains its own constant temperature, so birds burrow into it when the air temperature is high and thereby cool themselves. Conversely, they can warm themselves in the same way when the weather is very cool.
5. There will be no incident of breast blisters.
6. There will be no problem in cage layer fatigue.
7. Initial investment will be less when the land cost is low.

Disadvantages

- Housing density is lower than the cage system.
- There will be more feed wastage during spilling
- Litter-born diseases can occur, especially coccideosis, costing severe economic loss particularly in broiler industry.
- Disease spread faster due to mainly to free movements.
- Incidence of unclean or soiled eggs is higher.
- Birds consume more feed since they move about more freely wasting some energy; hence feed efficiency is inferior to the birds in cages.

5.2.3.2 Slatted or Wire Floor System

In this system of housing birds are reared on slatted or wire mesh floor where Slats- wooden pieces of 2.5-5 cm wide placed 2.5 cm apart which is running through the length of house. The slats placed 3 ft above the ground floor to allow accumulation of dropping. Feeding, watering & egg collection handled from outside the house only. Bird density can be 6–8 per square metre. This system of housing provides Cooler environment but expensive & suitable for adult bird only.

5.2.3.3 Combination of Slatted floor and Deep Litter

Here 60% is covered with slat area and 40% with litter area. Slats are constructed on either side of house against each side wall leaving central portion for litter floor. The total area is raised above the concrete floor by 0.5 metres or more to accumulate manure below the slatted area. It is very expensive and complicated.

5.2.3.4 Cage System Housing

Rearing of poultry on raised wire netting floor in smaller compartments, called cages. Initially, it was introduce for individual egg & pedigree recording & culling of poor layers but at present, 75% of commercial layers in the world are kept in cages. This system of housing is best suitable for keeping high density of birds, when space is a limitation. Scientific managemental practices can be followed in this system of housing. Feeders and waterers are attached to cages from outside, except nipple waterers, for which pipeline is installed through or above cages. Auto-operated feeding trolleys and egg collection belts can also be used .The droppings are either collected in trays underneath cages, on belts or on floor or deep pit under the cages.

Recommended Floor space

Chick (0 to 8 weeks) = 0.3 Sq.ft

Grower (9 to 16 weeks) = 0.5 Sq.ft

Layer (Above 17 weeks) = 0.6 Sq.ft

Types of cages

Based on Bird density: Single or individual bird cage, Multiple bird cage, Colony cage

Based on arrangement of cages: Battery type, Stair step cages (M and L type)

Based on Number of rows: Single deck, double deck, triple deck, four deck

Based on Bird type: Brooder, grower, layer, breeder

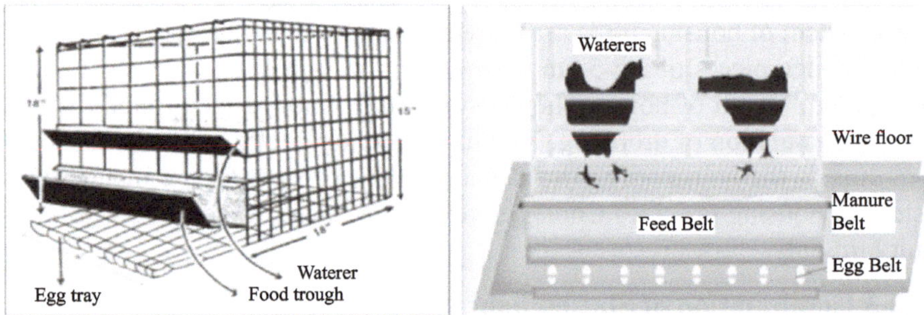

Figure 5.1: Structure of cages

California Housing (High Rise/ Elevated cage house)

This type of houses provides sufficient ventilation & waste management in tropical countries. The height of the shed is raised by 7-8 feet from ground level using concrete pillars. The distance between two pillars is 10 feet. Two feet wide concrete platforms are made over the pillars. The inter-platform distance is 5-8 feet depending upon the type of the cages used. The total height of the house is 20-25 feet and the width is 30-33 feet.

Figure 5.2: California Housing (High Rise/ Elevated cage house)

Advantages

- Easy egg collection
- Less problem due to cannibalism
- No nest requirement
- Record individual bird performance
- Better labor utilization
- Low parasitic infection
- Elimination of broodiness
- Avoidance of litter and its problems

Disadvantages

- More heat stress in summer
- Fly problem
- Incidence of broken egg is more
- Cage layer fatigue
- Fatty liver syndrome
- Poor shell quality
- Difficulty in manure disposal
- Breast blisters in case of broilers

Table 5.1. Cage specifications

	Front feeding length (inch)	Front and back height (inch)	Depth (inch)
Brooder cage	60	12	36
Grower cage	30	15	18
Conventional cage	15	18 & 15	18
Reverse cage	18	18 & 15	15

Following points to be considered while constructing shed

1. Shed should be located away from residential and industrial area
2. Basic amenities like water and electricity supply should be there
3. Proper ventilation should be provided
4. Marketing facilities should be considered

5. Proper road facilities should be there
6. Housing orientation should be towards east- west direction
7. Width of the house should not be more than 30 feet (for EC house it can be upto 40 feet)
8. Height of the sides from foundation to the roof line should be 8-10 feet (eaves height) and at the centre 10 -12 feet

5.3. Materials which are Used for Poultry Housing

5.3.1. Types of Roofing Material

5.3.1.1. Aluminium Sheet Roofing: Aluminium sheet roofing can be a good option for poultry house roofing because of its durability, resistance to corrosion and ease of installation. Aluminium sheet roofing is highly durable and can last for many years with proper maintenance.

Disadvantages

1. Aluminium is a highly conductive material, which means it can transfer heat and cold easily.
2. Aluminium is a good conductor of electricity, which can increase the risk of lightning strikes if the poultry house is located in an area prone to lightning storms.

5.3.1.2 Asbestos Sheets: Asbestos is a naturally occurring mineral that is resistant to heat and fire, making it an excellent material for use in high-temperature applications.

Disadvantages

1. Health hazards: Asbestos fibres are extremely hazardous to human health and can cause lung cancer, mesothelioma, and other respiratory diseases when inhaled. Exposure to asbestos fibres can occur during the installation, removal, or repair of asbestos-containing materials.
2. Environmental impact: Asbestos fibres are also harmful to the environment and can contaminate soil and water when disposed of improperly.

5.3.1.3. Thatched Roofs: It has been traditionally used for poultry houses in many parts of the world. Here are some advantages and disadvantages of using a thatched roof for a poultry house:

Advantages

1. Insulation: Thatched roofs provide good insulation, keeping the poultry house cool in hot weather and warm in cold weather. This can help to maintain a comfortable environment for the birds.
2. Natural material: Thatched roofs are made from natural materials, such as straw, grass, or reeds. This can make them a more sustainable choice compared to other roofing materials.
3. Low cost: Thatched roofs can be relatively low-cost compared to other roofing materials, especially if the materials are locally sourced.

Disadvantages

1. Fire hazard: Thatched roofs are a fire hazard, especially if the thatch is old and dry. They can easily catch fire from sparks or flames and can quickly spread to the rest of the building.
2. Maintenance: Thatched roofs require regular maintenance to prevent leaks, damage, and decay. This can be time-consuming and expensive, especially if the thatch needs to be replaced.
3. Pest problems: Thatched roofs can attract pests, such as birds, rodents, and insects. These pests can damage the thatch and create a nuisance for the poultry.

Types of roofing material

Aluminium Sheet Roofing

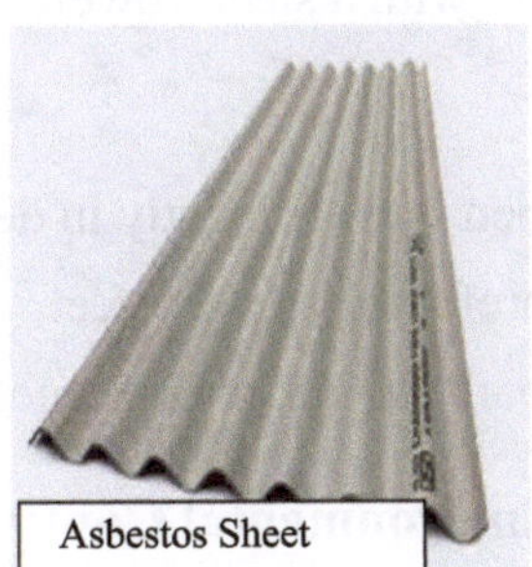

Asbestos Sheet

Thatched Sheet

5.3.2. Types of Floors

5.3.2.1. Katcha Floor: Also known as earthen floor, is a type of flooring used in poultry housing where the floor is made of natural soil.

Advantages

1. Low cost: Katcha floors are relatively inexpensive to construct compared to other flooring options.

2. Natural environment: Poultry feel more comfortable and natural in a katcha floor environment as it is similar to their natural habitat.

Disadvantages

1. Susceptible to erosion: Katcha floors are susceptible to erosion and can become uneven, which can lead to structural problems in the poultry house.
2. Attracts pests: Katcha floors can attract pests such as rodents and insects, which can cause damage to the poultry and the structure of the house.
3. Poor durability: Katcha floors have poor durability and may need to be replaced frequently,
4. Risk of disease: Katcha floors may increase the risk of disease transmission if they are not properly maintained and disinfected.

5.3.2.2. Concrete Floor

1. Concrete with rat proof device and free from dampness
2. Extended 1.5 feet outside the wall on all sides to prevent rat and snake problems
3. Consist of well-drained soil or gravel or concrete which is more desirable, it is easy to clean, durable and more rat proof
4. A concrete floor should be 80–100 mm thick and be made of a stiff 1:2:4 or 1:3:5 mix, laid on a firm base at least 150 mm above ground level, and given a smooth finish with a steel trowel

5.3.3. Doors

1. The door must be open outside mostly in deep- litter poultry houses
2. The size of door is preferably 6 x 2.5 feet
3. At the entry, a foot bath should be constructed to fill with a disinfectant

5.4. Climate-smart / Environmental Controlled Poultry House

In recent years, most poultry operation are intensive type houses with Environment Controlled house, in which inside conditions are maintained as near as to the bird's optimum requirements. A closed building with no windows, longitudinal preferably east to west, with big exhaust fans on west side while evaporative cooling pads on east side along with automatic feeding and drinking systems inside. Fully system is controlled with no manual controls, feeding system, watering system, manure collection system, egg collection system are all mechanized and automatic. ECH (Environment

Controlled House)helps to achieve better FCR, improving production, care of birds, control diseases and meet other safe breeding conditions and it helps to get one extra batch (or cycle) per year per shed.

5.4.1. Fans

Placed on the wind ward direction of the houses. Install slow speed, industrial fans 1m above the ground. Use 1x 620 mm rpm fan/1,000 layers in EC houses it is important to determine how much air flow through the building which determine the number of fans required.

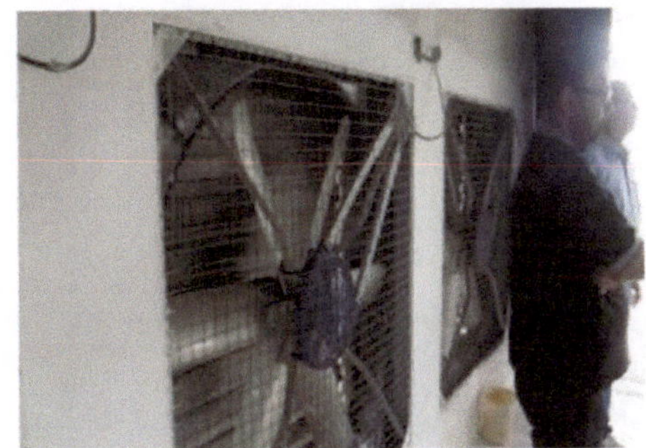

5.4.2. Insulation

A well-insulated building is needed for EC houses. It prevent condensation on the inside surfaces, reduce heat loss in cold weather, and reduce solar heat gain in warm weather

5.4.3. Evaporative Cooling Pads

Operate on the same cooling principle as foggers Cooled air entered the house, when it passes through the wet pads above which water runs through perforated pipes. This method avoids the problem of wet litter. Concrete coated bagasse pad, aspen fiber, rubberized hogshair pad and corrugated cellulose or fluted cardboard pads materials widely used as cooling pads.

5.4.4. Feeding

Automatic feeders have one loop of feeder chain and trough capable of feeding a given number of birds. A feeding control system to turn the feeders on and off. Feed monitoring systems are available to measure the amount of feed consumed by bird. The feed hopper in the house holds the feed before it goes out through the auger & into the feed pans. Feed goes through the auger lines & drops into each feed pan for the chickens to eat.

5.4.5. Egg and fecal Collection Conveyor Belt System

Belt is attached below the battery cage to collect egg and fecal collection.

6

Egg Production Practices Breeding and Feeding

Vidyasagar

Abstract

Poultry production in India has taken a quantum leap in the last four decades, emerging from use of unscientific farming practices to commercial production systems with state-of-the-art technological interventions. Today, the egg production system became highly specialised and more commercial in production practices. Changes in production practices have generated many technologies in basic disciplines like genetics, nutrition, management and physiology of birds. It's an area to which breeders have been giving increased attention to improve quality of the freshly laid egg.

6.1 Introduction

The egg production in the country has increased from 78.48 billion in 2014-15 to 114.38 billion in 2019-20. India ranks 3rd in egg production in the world. Annual growth rate of egg production was 4.99 per cent during 2014-15, there after there has been a significant improvement in the egg production with 10.19 per cent growth registered in 2019-20 over the previous year. The per capita availability of egg was 86 per annum in 2019-20. Less than a generation ago the main supply of eggs was from small backyard unit. Processors, handlers, and distributors of eggs and egg products need an understanding of egg-production practices to cope more intelligently with problems of product quality and character. In same way, food scientist involved with quality control, new product development and composition studies can benefit from the knowledge how egg is produced. Therefore the purpose of this chapter is to briefly sketch about egg- production practices.

6.2 Breeding practices

1. Under breeding practices irrespective of the fact whether it is broiler or layer breeder we are having following types of breeding flocks.
2. Pure lines (maintained by foundation/principal breeder)
3. Grandparent stock (maintained by franchise hatcheries)
4. Parent stock (maintained by associate hatcheries)
5. Commercials (maintained by poultry farmers)
6. Leghorn is the breed which is normally used for egg purpose

6.2.1 Commercial Layer Strains

1. Babcock : B300 (70% of Indian share), B380 (brown egg layer)
2. Hendrix : Bovans white, Bovans brown
3. Hy-line: W-77, W-36, W-88, Hy-line brown, Hy-line silver brown (popular strain in world)
4. Lohmann: Lohman white, Lohman brown Eubrid, H&N, Dekalb etc.

6.3 Feeding Practices

6.3.1 General Principles of Feeding in Poultry

1. Poultry feed should contain all essential nutrients like protein, fat, carbohydrates, energy, fibre, minerals, vitamins & moisture in proper proportion depending on type, category of bird & season.
2. The feed should be free from all pathogenic organisms like salmonella, E coli, etc & also devoid of toxins like gossypol, aflatoxin, etc.
3. The finished feed should not be stocked for more than 1 to 1.5 months to avoid loss of nutrient & to prevent development of rancidity, fungal growth, moulds & also spoilage by rodents.
4. Formation of cakes in feeders should be avoided to stop the growth of fungus & moulds in feeders.
5. Feeders should not be filled more than 1/3rd to ½ level to control wastage.
6. The nutrient level should be changed as per need of season.
7. Minimum two feedings in the form of all mash or pellets are good for optimum consumption & to ensure correct intake of micronutrients.
8. Poultry birds at any stage should not be under or over fed.

6.3.2 Methods of Feeding

Whole grain feeding method

This is the traditional practice of feeding in the backyard poultry in villages. The birds are allowed free roaming and grains are fed at home. As poultry are reared at present on commercial lines, this system has no relevance. However, with the craze for organic 'natural' eggs now, this system of feeding is coming back.

Grain & mash feeding method

Mash means a mixture of grounded feedstuffs. Whole grain feeding is supplemented with high protein mash mixture to provide additional feeding. Mash also helps to provide vitamins and minerals that may be deficient in the all grain feeding system. This is not a common practice, but can be used for improving feeding levels of backyard poultry.

All mash feeding method

This is the most common system practiced at present. This comprises of a mixture of ground grains, millet feeds, and protein and mineral/vitamin supplements combined in calculated proportions to meet the nutrient requirements of the birds. It is an all-in- one type of complete feed. Different mashes with different protein and energy levels are prepared for very young chicks – starter mash, for growing birds – grower mash, for layers – layer mash and for layers and broiler mash for broiler birds are prepared. Mash is the common method of feeding large-scale commercial poultry complexes. If stored for long period occurrence of mycotoxicosis is the main disadvantage in this type of feed.

Pellet feeding method

Pellets are made from mash, which is then heated and compressed into a hard compact pellet. Chicken feed pellets are designed to be a complete feed, with the right levels of proteins, vitamins and minerals. Because pellets are larger and more difficult to digest, they are generally used for adult hens and not for young chicks or pellets. As the feed is heat treated occurrence of mycotoxicosis can be avoided.

Crumble form of feeding method

Crumbles are made from whole pellets, which are cracked or rolled into a smaller size. Chicken Crumble is designed to be a complete feed, with the right levels of proteins, vitamins and minerals. Crumble is a popular choice for pullets (teenage chickens) as a transition from mash to pellets. However, crumble is also used as feed for chicks and laying hens.

Figure 6.1: Mash Feed

Figure 6.2: Crumbles Feed

Figure 6.3: Whole grains Feed

Figure 6.4: Pelletes Feed

6.3.3 Feeding Systems in Poultry

For layers restricted method of feeding is normally practiced

1. Controlled feeding/ Restricted feeding
2. Supplementary feedings

6.3.4 Methods of Ration Restriction

1. Skip-a-day method: not fed for one day in a week.
2. Feeding limited quantity every day: measured feed is given once a day usually in afternoon. This is the best method of ration restriction.
3. Reducing feeding time: by increasing dark hours or taking off feed.
4. Limiting nutrient intake: protein & energy intake is limited by diluting the feed with more fibre content.

Advantages of restricted feedings

1. Delays sexual maturity & onset of egg production by few days to 3 or 4 weeks depending on level of restriction.
2. Reduces body weight of bird at sexual maturity usually by reducing the amount of fat deposition.
3. Reduces cost of growing pullet & results in better liveability during egg production.
4. Increases the size of initial eggs laid as age is an important factor to regulate the size of eggs.
5. Careful feed restrictions result into overall better economy with birds fed ad libitum.

6.3.5 Supplementary Feeding

Improving the nutritional quality of poultry feed by supplementing minerals, vitamins, and amino acids are essential to maintaining a healthy, valuable flock. There are three main groups of nutritional supplements that can be added to poultry feed: proteins and amino acids, vitamins, and minerals. These groups can be broken down further to come up with the five most common types of poultry feed supplements: Electrolytes, Amino acids, Vitamins, Minerals, Probiotics and fermentation products.Using a healthy balance of poultry feed additives and supplements will ensure your flock can thrive and bring production levels back to optimal levels.

Table 6.1: Layer feed specification

S.No.	Characteristics	Layer feed			
		CFL	GFL	LFP-I	LFP-II
1	Moisture (% by mass) *max.*	11	11	11	11
2	Ether extract (% by mass) *min.*	2	2	2	2
3	CP (% by mass) *min.*	20	16	18	16
4	CF (% by mass) *max.*	7	9	9	10
5	Acid insoluble ash (% by mass) *max.*	4	4	4	4.5
6	salt (% by mass) *max.*	0.5	0.5	0.5	0.5
7	Metabolizable energy (kcal/kg)*min.*	2800	2500	2600	2400
8	Calcium (% by mass) *min.*	1	1	3	3.5
9	Phosphorous (% by mass) *min.*	0.7	0.65	0.65	0.65
10	Methionine (% by mass) *min.*	0.4	0.35	0.35	0.3
11	Lysine (% by mass) *min.*	1	0.7	0.7	0.65
12	Aflatoxin B1 (ppb) *max.*	20	20	20	20
13	Methionine + cystine (% by mass) *min.*	0.7	0.6	0.6	0.55

6.3.6 Feeding of Layers (As per BIS 2007)

1. Chick feed for layer (CFL)- a ration to be fed to chicks. intended for egg production, from 0 to 8 weeks.
2. Grower feed for layer (GFL)- a ration to be fed to growing chickens, intended for egg production, from 9 to 20 weeks or until laying commences.
3. Layer feed for phase–I (LFP-I) - a ration to be fed to laying birds from 21 weeks to 45 weeks.
4. Layer feed for phase –II (LFP-II) - a ration to be fed to laying birds from 46 weeks to 72 weeks.

7

Nutritive Importance of Eggs

Kiran M

Abstract

Egg is an enclosed source of all the and micronutrients necessary for embryonic development till hatching. The egg's excellent balance and range of nutrients, together with its great digestion and low cost, have elevated it to the status of a fundamental human diet. Eggs continue to be a very nutritious dietary product for humans, including the elderly and children, and are widely consumed across the world. In addition, there is persuasive evidence that eggs contain several undiscovered bioactive chemicals that may be of great use for preventing/treating illness.

7.1 Introduction

From a nutritional standpoint, eggs are particularly interesting since they include important fats, proteins, vitamins, minerals, and trace elements. They also provide a moderate calorie supply (about 140 kcal/100 g), have a wide range of culinary applications, and are relatively inexpensive. Indeed, it has been shown that eggs are the second-cheapest animal source for zinc and calcium and the lowest-cost source of proteins, vitamin A, iron, vitamin B12, riboflavin, and choline. Eggs include a wide variety of physiologically active ingredients in addition to offering babies and adults well-balanced nutritional needs. These elements are distributed throughout the many internal egg components. It must be noted that while eggshell membranes are edible, eggshell and its closely related eggshell membranes are often not ingested (Figure 7.1). There are around 3 billion chickens worldwide that have been bred exclusively to produce table eggs for human consumption without fertilisation. Although a limited proportion of egg proteins are not digested, especially when egg is ingested as a raw foodstuff, egg components are also said to be highly digestible. The greater digestibility of cooked egg proteins is due to structural protein denaturation brought on by heating, which makes it easier for digestive enzymes to function hydrolytically.

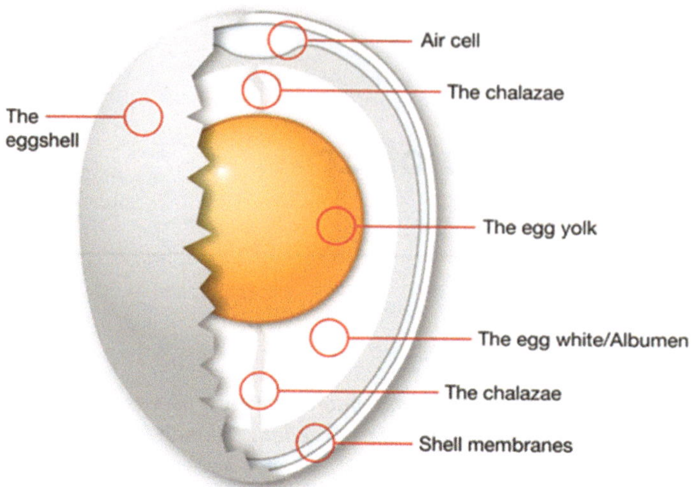

Figure 7.1: Egg structure

Source: Nys, Y., & Guyot, N. (2011). Egg formation and chemistry. In Improving the safety and quality of eggs and egg products (pp. 83-132). Woodhead Publishing Ltd.

7.2 Egg Nutrients

Egg white and yolk have equal amounts of egg proteins, whereas the yolk mostly contains lipids, vitamins, and minerals (Table 7.1). The majority of an egg is made up of water, and it's important to notice that there are no fibres in the egg. When primary elements like water, proteins, lipids, and carbohydrates are taken into account, the relative concentration of egg minerals, vitamins, or certain fatty acids may differ from one national reference to another while being internationally similar. In contrast to minor components, which are influenced by a variety of circumstances, including hen nutrition, the key egg nutrients are in fact highly stable and depend on the egg white to yolk ratio. Water, protein, fat, carbs, and ash make up around 76.1%, 12.6%, 9.5%, 0.7%, and 1.1% of a complete, uncooked, and freshly hatched egg, respectively. In **Table 7.2,** the precise proximate composition is presented.

Table 7.1: Variation in egg yolk and albumin nutrent composition

Composition	Albumen	Yolk
Water	87.72	55.02
Crude protein	10.82	15.50
Carbohydrate	0.85	1.09
Fat	0.19	26.71
Ash	0.42	1.68

(*Source:* USDA National Nutrient Database for Standard Reference, Release 23 (2010))

Table 7.2: Proximate composition of raw, fresh eggs of different poultry species (per 100 g, edible portion)

Composition	Unit	Duck	Goose	Quail	Turkey	Chicken
Water	g	70.83	70.43	74.35	72.5	76.15
Energy	kcal	185	185	158	171	143
Protein	g	12.81	13.87	13.05	13.68	12.56
Fat	g	13.77	13.27	11.09	11.88	9.51
Ash	g	1.14	1.08	1.1	0.79	1.06
Carbohydrate	g	1.45	1.35	0.41	1.15	0.72
Fibre	g	0	0	0	0	0
Sugar	g	0.93	0.94	0.4	NA	0.37

(*Source:* USDA National Nutrient Database for Standard Reference, Release 23 (2010))

7.2.1 Proteins in Eggs

Protein-rich eggs are a healthy option. Both the egg white and the egg yolk contain a large number of proteins. During the course of embryonic development, hundreds of proteins are expressed and each one has a unique physiological purpose. A protein's value is determined mostly by how well it is absorbed and how many important amino acids it contains. High-quality proteins are those that are easily digestible and contain all the necessary amino acids at concentrations higher than the FAO/WHO/UNU standard values (FAO/WHO/UNU, 1985). Out of all the main dietary proteins, egg protein has the highest digestibility (97%). When compared to other protein sources, table eggs offer the greatest nutritional quality protein due to their complete amino acid profile and abundance of important amino acids that nearly matches human needs (FAO protein value = 100). Therefore, all other proteins are measured against egg protein because of its status as the gold standard of nutrition.

About 27% of an adult's RDI for protein may be found in a single big egg (6 g total; 3.6 g white and 2.4 g yolk). Due to the limited protein digestibility and poor biological protein value of many plant proteins, ovovegetarians may benefit greatly from include egg proteins in their diets. Due to their good essential amino acid profile, high digestibility, and simplicity of preparation, egg proteins are also a suitable protein source for youngsters and the elderly.

Protein content in a raw, fresh egg is 12.5 grammes per 100 grammes on average, with the egg yolk and vitelline membrane containing 15.9 grammes and the egg white containing 10.90 grammes per 100 grammes. Both the egg white and the egg yolk are very protein-rich. During the course of embryonic development, hundreds of proteins are expressed and each one has a unique physiological purpose. This separate tissue composition of the egg yolk and white provides an explanation for the compartment specificity of some proteins. To a large extent, the liver is responsible for the production of egg yolk, whereas the oviduct is responsible for the synthesis and secretion of egg white following the ovulation of the mature yolk.

7.2.2. Egg Lipids

About 60% of the dry weight of an egg yolk is made up of lipids, the yolk's primary component. The lipids in fresh eggs are mostly bound to protein in the form of lipoprotein particles, and to a lesser amount to carbohydrates. Triglycerides, phospholipids, cholesterol, cerebroside, and a few more small lipids make up the lipids. Triglycerides in egg yolk are predominantly composed of oleic acid (C18:1) and palmitic acid (C16:0). Other important fatty acids include linoleic acid (C18:2) and stearic acid (C18:0). Egg yolk's ratio of omega-6 to omega-3 fatty acids is easily modifiable by diet. The egg yolk is a popular ingredient in both commercial and residential kitchens due to its effectiveness as a natural emulsifier. There are mostly phospholipids in the egg yolk's lipid fraction, which makes up around 33% of the yolk's total weight. Phosphatidylcholine accounts for close to 80%, phosphotidylethanolamine for 11.7%, and sphingomyelin and lysophosphotidylcholine for around 2% each. For proper cell and brain growth, as well as phospholipid production in cell membranes, neurotransmission, transmembrane signalling, and lipid and cholesterol transport and metabolism, choline and its derivatives are required. Bioavailable choline may be found in eggs.

Table 7.3. Lipid composition of raw, fresh eggs of different poultry species (per 100 g, edible portion)

Composition	Unit	Duck	Goose	Quail	Turkey	Chicken
Fatty acids, total saturated	g	3.681	3.595	3.557	3.632	3.126
Fatty acids, total monounsaturated	g	6.525	5.747	4.324	4.571	3.658
Fatty acids, total polyunsaturated	g	1.223	1.672	1.324	1.658	1.911
Cholesterol	mg	884	852	844	933	372

(*Source*: USDA National Nutrient Database for Standard Reference, Release 23 (2010))

The majority of the sterol in egg yolk is cholesterol. In uncooked eggs, cholesterol accounts for 1.6% of the yolk and 5% of the lipids. Around 84% of total cholesterol is free cholesterol, whereas the remaining 16% is cholesterol ester. Cholesterol in egg yolks is mostly derived from the feed that the hens eat, but part of it is also created in the hen's body during the process of forming the egg yolk. Cholesterol, as said, is a vital part of cell membranes and a precursor to hormones, vitamin D, and bile acids. Glycolipids are what cerebrosides fall under. They are made up of sphingosine, a sugar (either galactose or sucrose), and a nitrogen-containing base. The fatty acid composition of two cerebrosides (ovophrenosin and ovokerasin) isolated from egg yolk is different.

7.2.3. Carbohydrates in Eggs

Egg does not include any fibres, and it only has a very little amount of carbs (0.7% of its total). The carbohydrate content of an egg is split between the egg yolk and the egg white. Glucose is the primary form of free sugar found in eggs (with an average of 0.37 grammes per 100 grammes of whole egg), and it is mostly found in egg white (0.34 g per 100 g of egg white versus 0.18 g per 100 g of egg yolk). Both the raw egg white and the raw egg yolk have been shown to contain minute levels of the sugars fructose, lactose, maltose, and galactose. Egg proteins are predominantly composed of glycoproteins, which go through a series of post-translational glycosylations before being secreted by the reproductive tissues of the hen to form the egg's yolk, membranes, and white. Because of this, carbohydrates have a significant presence in egg proteins.

7.2.4. Vitamins and Minerals in Eggs

Except for vitamin C, eggs contain the majority of vitamins essential for human nutrition. The egg yolk is the only source of the fat-soluble vitamins. One egg can provide about 11% of the recommended daily allowances for vitamin A, vitamin D, riboflavin, and panthotenic acid in the United States. There are two dietary forms of vitamin A: preformed vitamin A (retinol) and provitamin A. (beta-carotene). In Western nations, the most prevalent source of vitamin A is preformed vitamins found in animal products and egg yolk. Vitamin A is essential for regulating development, cell repair, and differentiation, as well as maintaining healthy skin and eyes. The vitamin content is influenced by genetics, egg production rate, and the nutrition of the hen.

Eggs are a major mineral source. Iron is mostly associated to the proteins ovotransferrin (in egg white) and phosphivitn (in egg yolk) in eggs, which can restrict its bioavailability. Eggs contain selenium as selenomethionine, which has a greater bioavailability than inorganic selenate or selenite. Selenium with vitamin E form an important antioxidant enzyme (glutathione peroxidise).

Table 7.4 Mineral composition of raw, fresh eggs of different poultry species (per 100 g, edible portion)

Composition	Unit	Duck	Goose	Quail	Turkey	Chicken
Calcium, Ca	Mg	64	60	64	99	56
Iron, Fe	Mg	3.85	3.64	3.65	4.1	1.75
Magnesium, Mg	Mg	17	16	13	13	12
Phosphorus, P	Mg	220	208	226	170	198
Potassium, K	Mg	222	210	132	142	138
Sodium, Na	Mg	146	138	141	151	142
Zinc, Zn	Mg	1.41	1.33	1.47	1.58	1.29
Copper, Cu	Mg	0.062	0.062	0.062	0.062	0.072
Manganese, Mn	Mg	0.038	0.038	0.038	0.038	0.028
Selenium, Se	Mcg	36.4	36.9	32	34.3	1.1

(*Source:* USDA National Nutrient Database for Standard Reference, Release 23 (2010))

Table 7.5: Vitamin composition of raw, fresh eggs of different poultry species (per 100 g, edible portion)

Composition	Unit	Duck	Goose	Quail	Turkey	Chicken
Vitamin C/ ascorbic acid	Mg	0	0	0	0	0
Thiamine	Mg	0.156	0.147	0.13	0.11	0.04
Riboflavin	Mg	0.404	0.382	0.79	0.47	0.457
Niacin	Mg	0.2	0.189	0.15	0.024	0.075
Pantothenic acid	Mg	1.862	1.759	1.761	1.889	1.533
Vitamin B-6	Mg	0.25	0.236	0.15	0.131	0.17
Folate, total	Mcg	80	76	66	71	47
Choline, total	Mg	263.4	263.4	263.4	NA	251.1
Vitamin B-12	Mcg	5.4	5.1	1.58	1.69	0.89
Vitamin A, IU	IU	674	650	543	554	540
Vitamin E	Mg	1.34	1.29	1.08	NA	1.05
Vitamin D	IU	69	66	55	NA	82
Vitamin K (phylloquinone)	Mcg	0.4	0.4	0.3	NA	0.3

(*Source:* USDA National Nutrient Database for Standard Reference, Release 23 (2010))

Table 7.6 Amino acid composition of raw, fresh eggs of different poultry species (per 100 g, edible portion)

Composition	Unit	Duck	Goose	Quail	Turkey	Chicken
Tryptophan	g	0.26	0.282	0.209	0.219	0.167
Threonine	g	0.736	0.797	0.641	0.672	0.556
Isoleucine	g	0.598	0.647	0.816	0.855	0.671
Leucine	g	1.097	1.188	1.146	1.201	1.086
Lysine	g	0.951	1.03	0.881	0.924	0.912
Methionine	g	0.576	0.624	0.421	0.442	0.38
Cystine	g	0.285	0.309	0.311	0.326	0.272
Phenylalanine	g	0.84	0.91	0.737	0.773	0.68
Tyrosine	g	0.613	0.664	0.543	0.569	0.499
Valine	g	0.885	0.958	0.94	0.985	0.858
Arginine	g	0.765	0.828	0.835	0.876	0.82
Histidine	g	0.32	0.346	0.315	0.33	0.309
Alanine	g	0.631	0.683	0.762	0.799	0.735
Aspartic acid	g	0.777	0.841	1.294	1.357	1.329
Glutamic acid	g	1.789	1.937	1.662	1.742	1.673
Glycine	g	0.422	0.457	0.434	0.455	0.432
Proline	g	0.48	0.52	0.518	0.543	0.512
Serine	g	0.963	1.043	0.992	1.04	0.971

(*Source:* USDA National Nutrient Database for Standard Reference, Release 23 (2010))

7.3. Antinutritional Factors in Eggs

Egg's primary proteins contain protease inhibitors, which hinder digestive enzymes such as pepsin, trypsin, and chymotrypsin, delaying the normal breakdown of egg proteins. Egg white is a significant source of ovostatin, ovomucoid, ovoinhibitor, and cystatin. In addition, a few of these compounds (ovoinhibitor, ovomucoid, and cystatin) have many disulfide bonds that presumably give modest resistance to denaturation by proteases and stomach acids. Some of these antinutritional substances may be partially denatured during the cooking process, hence increasing the access of digestive proteases to proteins. In addition, the high concentration of vitamin-binding proteins in egg may further restrict vitamin absorption; the avidin that binds vitamin B12 (biotin) has the greatest known affinity in nature between a ligand and a protein. The tight complex formed between avidin and its bound vitamin B8 may diminish the bioavailability of biotin for consumers.

7.4. Eggs from Alternative Poultry Species

Eggs of the quail are a rich source of beneficial trace elements and vitamins. Their nutritious content is three to four times that of chicken eggs. They contain 13% protein, which is greater than the 11% found in chicken eggs. Compared to chicken eggs, quail eggs contain twice as much vitamin A and vitamin B2 and contain twice as much vitamin B1. In addition, quail eggs contain five times more iron and potassium than chicken eggs. Additionally, they contain more phosphorus and calcium. Due to their remarkable nutritional value, quail eggs are regarded a diet food. Due to the absence of bad cholesterol (LDL) and the abundance of good cholesterol (HDL), even elderly persons can consume quail eggs. Asthma, gastritis, stomach ulcer, duodenal ulcer, etc., are believed to be treatable with quail eggs, according to a number of studies. Also considered to have aphrodisiac powers are quail eggs.

About 15% of a person's daily fat needs may be met by eating one fried duck egg. The 9 grammes of protein in a duck egg is comparable to 18% of the daily requirement. As a general rule, duck eggs are smaller than goose eggs but considerably larger than chicken eggs. They're essential in many places of the world due to their great nutritional value. The health advantages of eggs can be enjoyed by those who are allergic to chicken eggs because ducks also lay eggs. Duck eggs are more nutrient rich than chicken eggs because they contain less water overall. All nine of the body's required amino acids may be found in duck eggs, making them a complete protein source. Rubbery and membrane-tough, duck egg shells are surprisingly tough. Due to this, duck eggs may be safely transported across vast distances. The richness of the baked goods can be attributed to the yolk, which has a distinct orange hue.

Compared to a chicken egg, a goose egg is a whopping three times larger. Goose eggs have a much tougher shell and a more robust taste than regular chicken eggs. Due to their bigger size, many of the nutrients found in chicken eggs are increased. The USDA daily value for calcium, vitamin A, and iron found in a single goose egg is 9%, 19%, and 29%, respectively. It has 3 times as much selenium as an egg, or 53.1 mg. The pigment lutein, which helps maintain healthy eyes and skin, is abundant in goose eggs. Choline, one of the B-complex vitamins, may be found in a goose egg with a concentration of 379 milligrammes. Choline has a recommended dietary allowance (RDA) of 425 milligrammes for most adult women and 550 milligrammes for most adult men. Choline is essential for cell growth and communication. Liver disease, atherosclerosis, and brain dysfunction have all been linked to insufficient choline levels.

Turkey eggs, in comparison to chicken eggs, are significantly bigger and have a larger shell with a reddish pattern on them. The exterior and membrane are more robust, making them difficult to penetrate. Turkey eggs have a similar flavour and nutritional profile to chicken eggs, but they come in much larger volumes. Turkey eggs are almost 50% bigger than chicken eggs, although they have the same nutritious profile. Eggs from a turkey can be eaten; they make delicious omelettes and baked goods. Three times as much cholesterol as in a chicken egg may be found in a turkey egg, and far above the daily limit of 300 milligrammes. People with hypercholesterolemia should avoid eating turkey eggs as a result. Since turkey hens require more food than chicken hens and lay fewer eggs per bird, turkey eggs are not mass-produced.

8

Preservation of Eggs

Wilfred Ruban S and Kiran M

Abstract

Eggs are a highly nutritious food that contains proteins, minerals, fats, iron, phosphorus vitamins (A, B, D, E, and K) needed by human beings. The fully mixed egg contains about 65% water, 12% proteins, and 11% fat. These various nutrient contents present in eggs make it an excellent source for bacterial microflora, including pathogenic bacteria. Thus, various preservation methods are used to eliminate the growth of spoilage-causing microorganisms.

8.1 Introduction

Egg is a perishable product having short shelf life. The table eggs are of highest quality when they are laid. Preservation of eggs should start from the point of production itself.

The following practices are recommended as routine for the production of quality eggs on the farm.

1. Collection of eggs at least 3 times daily.
2. Using a clean receptacle with ventilated sides and bottom, preferably filler flats.
3. Careful handling of eggs during collection and while storing in filler flats etc.
4. Cooling the eggs quickly to 50°F or less at 75-85% relative humidity.
5. Marketing the eggs at least twice weekly.

However, the following changes on storage bring about spoilage.

1. Moisture loss through the pores
2. Gaseous (CO_2) loss
3. Change of pH
4. Movement of water from thick albumen to other albumen layers and to yolk

5. Entry of microbes through the pores
6. Embryonic development
7. Absorption of foreign odours

Egg preservation is therefore aimed at minimizing these changes in eggs under storage by employing various methods. Eggs can be preserved to improve the keeping quality either as shell eggs or as egg liquid (whole egg or yolk or albumen).

8.2 Shell Egg Preservation

The methods for preservation of egg shell are based on simple principle of retarding the microbial growth and sealing the pores of the shell to minimize the evaporation of moisture and escape of gases, thus reducing the physico-chemical changes in the egg. Sometimes, a combination of methods is employed for effective preservation.

8.2.1 Egg Cleaning

8.2.1.1. Dry Cleaning

This was the recommended practice before the development of sanitizing compounds capable of maintaining washing water and equipment in a condition so that wet cleaning methods could be used. The simplest dry cleaners are hand abrasive blocks to scratch the dirt or stain from the shell surface. Abrasives were put on mechanically rotated wheels so that eggs could be rapidly cleaned by holding individual eggs against the rotating abrasive surface. Although mechanical equipment for dry cleaning shell eggs was developed it has been made obsolete by wet cleaning procedures. Disadvantage of dry cleaning is weakness of the shell. Only dirty and unsound eggs should be cleaned.

8.2.1.2. Wet Cleaning

For clean appearance of eggs, washing is more effective and simplest method for removing dirt and stains from the shell surface. Egg wash water temperature need to be at least 11°C higher than egg temperature. The temperature differential between the eggs and wash water also critical, as a difference of more than 10° C results in increased number of thermal cracksdue to expansion of contents of egg. In older eggs with enlarged air cells, the thermal checking is not a serious problem. Also a water temperature of at least of 35°C (95°F) is needed for adequate cleaning. By not recycling water and by giving careful attention to the temperature difference between eggs and wash water eggs can cleaned with a minimum of hazard to quality.

There is possibility of bacterial penetration of the shells, with resultant rotten eggs. A temperature difference of more than 10° C results in an increased number of thermal cracks due to expansion of contents of the egg. Cold eggs subject to hot water will expand too fast and crack the shell (expansion checks) on the other hand the interior contents of eggs that have been washed in hot water and then rinsed by immersion in cool water contract and aspirate bacteria through shell. The organisms multiply and the eggs spoil. Modern egg washers are designed to minimize the above conditions by spraying water, rather than immersing them using a sanitizer in water along with detergent for cleaning them using rinse water warmer than the wash water and finally dry the eggs with hot air. A general recommendation is to handle shell eggs at less than10° C at all times.

Modern egg washers are designed to minimize the above conditions by spraying the eggs with water, rather than immersing them, using a sanitizer in the water along with detergent for cleaning them using rinse water warmer than the wash water, and finally dry the eggs with hot air.

8.2.2 Oil Treatment

Oil coating preserves the egg by forming a thin film on the surface of the shell and thereby sealing the pores. The treatment should preferably be given within a few hours of lay to retain better internal quality. Washing and subsequent coating preserve eggs much better than coating alone. The oil used must be odourless, colourless and free from fluorescent materials. Light mineral oil of food quality are normal used. Vegetable oil such as groundnut oil mixed 0.0125% BHT is a good sealing agent. Mineral oils of food grade are less susceptible to oxidative changes during storage. Eggs can either dipped in oil or sprayed with it.

8.2.3 Lime Treatment

Under village conditions liquids like lime water and water glass are very use full. This is old method of preserving the quality of shell eggs especially for home use. In lime sealing, one liter of boiling water is added to one kg of quick lime. The fluid is brought to room temperature, and 4-5 liters of cold water, and 250g of table salt are added to it. On the mixture setting down the solution is strained. The eggs are immersed in thin clear fluid for 16-18 hours, and then taken out. Eggs are dried at room temperature and transferred to filler flats. The eggs can be stored for 3 to 4 weeks at room temperature. The preservative effect of the lime water is partially due to its alkalinity. It deposits a thin film of calcium carbonate on the egg shell and partially seal the pores. The water glass is prepared by diluting 1 part of sodium silicate with 10 parts, of water and the eggs are left immersed over right in water glass. It deposits a thin precipitate of silica on the surface of eggshell, possesses intrinsic antiseptic properties and dose not impart odour or taste to the eggs. When started at 10 to 30 °C the water glass preserved eggs were comparable to fresh eggs in all attributes. Egg can be stored at 13 to 15° C for six months.

8.2.4 Thermal Processing

This method is good for fertile eggs, since it kills embryos and therefore is also known as 'defertilization' method. This includes flash heat treatment, thermostabilization and simultaneous coating.

8.2.4.1. Flash Heat Treatment

Flash heat treatment was introduced by Romanoff and Romanoff (1944) based on the immersion of egg in boiling water for 5 seconds. This resulted in egg with superior storage characteristics at storage temperature of either 5° C or 21° C. Immersion for only 3 seconds in boiling water was effective in reducing bacterial spoilage in fresh eggs.

8.2.4.2. Thermostabilization

A quite different treatment for the preservation of shell eggs was reportedly Funk in 1943. This involved for immersion in hot water or oil for sufficient time tokill the embryo in fertile eggs and to stabilize the thick albumen. The time and temperature suggested by Funk were, immersion in water and oil at 60° C for 14minutes. Simultaneous oil treatment and thermostabilization complement each other in maintaining inner quality of egg. The problems of commercial application of thermostabilization are 1) the coagulation temperature of egg white varies with the pH of the egg white. 2) temperature achieved by egg is influenced by egg size, starting temperature of egg, agitation of heating medium, 3) as well as time and temperature.

8.2.5 Cold Storage

For start period storage, fresh eggs should be 12.5-15.5 °C and 70 - 80% relative humidity. For long term storage the room temperature should be at -10° C and relative humidity 80-90%, as this relative humidity will sufficiently retard evaporation without danger of mould growth. Eggs can be oil treated prior to cold storage to enhance their keeping quality. The quality of the shell eggs can be maintained for 6 months in a cold storage. A final consideration in the presentation of egg quality is the maintenance of flavor, eggs will pick up flavours from storage areas, such as onion, garlic, apples, oranges, decaying vegetable matter, oil, gasoline or organic solvent. Oil dipped eggs could be kept at 14° C and 90% RH, or 13° C and RH 85% or 20° C and RH 72% for 8 months, 60days and 28 days respectively against 6 months, 30 days, and 7 days at corresponding storage temperatures of un oiled eggs.

8.2.6 Pickling

The eggs can be preserved by pickling also. It can be done with eggs of small size like that of quail eggs as it will allow proper penetration of pickling solution throughout the egg. The Central Avian Research Institute(CARI) has developed a technique for pickling of quail eggs.

8.2.7 Overwrapping

For over wrapping of eggs polyethylene, cellophane, polyvinylidene and other transparent, thin but sufficiently strong, films are used. These films should be impervious to gases and moisture. Over wrapping of eggs in different atmosphere like carbon dioxide, vacuum etc. have been tried.

8.2.8 Ultrasound Method

Ultrasonic is a high-power sound wave at frequencies between 16 kHz and 100 MHz. In this method, the sonic wave is passed through the egg liquid and the changes occur in the pressure which leads to cavitation, which causes gas bubbles in the liquid causing a bactericidal effect.

8.2.9 Microwave Method

Microwaves used in the food industry for heating are of frequencies ranges from 300 MHz to 300 GHz. Microwave heating is a method in which electromagnetic waves are used to generate heat in food. Electromagnetic waves can reduce Salmonella enteritidis, which is often found in shell eggs.

8.2.10 High-Pressure Process (HPP)

It is a non – thermal pasteurization process in which food is subjected to high pressure in the region of 3300 -600 MPa for about 10 minutes. The application of 600 Mpa for a 2 minutes cycle in boiled eggs was able to extend the shelf life of these products during refrigeration. HHP at a pressure between 200 and 350 MPa can be used for liquid eggs as did not cause detectable protein denaturation.

8.2.11 Pulse Electric Field (PEF)

The pulsed electric field is one of the non-thermal food preservation technologies in which food is subjected to short pulses (1-100 μs) of high electric fields with a duration of nano to milliseconds and intensity of 10 – 80 kV/cm to foods placed between two electrodes. The application of PEF in egg products has been found to control spoilage or pathogenic microorganisms.

8.2.12 Ozone

O_3 is a triatomic form of oxygen that has been gaining space in food processing due to its high sanitizing power and is also known as a highly reactive antimicrobial agent. O_3 between 4 and 6 $mg.L^{-1}$ could be used to maintain the internal quality of the eggs and extend their shelf life.

8.3 Liquid Egg Preservation

8.3.1 Egg Freezing

The freezing of the internal contents of eggs is now a common method of preservation specially in developed countries. The eggs are first candled and when they are broken out, the smell and appearance of the contents are noted for any possible defects. The yolk and the white may be frozen separately with addition of 5% glycerine. The egg contents are then freezed in 30-40 lb., tin at a low temperature range of 10ºF to 30ºF below zero. The contents are then kept at a low temperature until required for use. In case the storage temperature is zero or below, the frozen eggs may be stored with little or no loss of flavour for 12 months or longer.

8.3.2 Egg Drying

Egg drying is now largely practiced in place of freezing. Although the process is more expensive but there is a considerable saving in transport and less need for cold storage. The egg contents are dried at a temperature of 1600F and stored less than 500F to convert white, yolk or the whole egg into a fine powder. The whole egg is of use for bakery products, the yolk for flours and the albumen for confectionaries.

9

Grading of Eggs

Wilfred Ruban S and Kiran M

Abstract

The egg business uses egg grading to cut down on wastage and consumer uncertainty about egg quality. With standardised grading systems in place, the egg industry can communicate with customers and other businesses on a common platform, and shoppers can be certain that their eggs will always be of high quality. For poultry shows, the grading of eggs is just as crucial. Egg judging standards, guidelines, and instructions for competitions are discussed in this chapter.

9.1 Introduction

Many cultures throughout the world see the egg as nature's perfect meal. Many other kinds of animals lay eggs, yet the vast majority of people eat chicken eggs. There are a number of reasons why the quality of an egg matters today. Grading eggs is so essential. Quality, size, shape, weight, and other characteristics are all taken into account while sorting eggs for grading. To put it simply, food quality refers to the traits of a product that determine whether or not it is accepted by the target audience. Egg quality is based on both industry standards and consumer expectations. Considerations from the outside include egg shell composition, hue, form, and feel. Air cells, egg white, egg yolk, etc., are all examples of internal quality.

When describing a dish, quality refers to all of the factors that go into making it desirable to eat. Good, better, best, or C, B, and A grades are used to categorise a commodity into distinct quality categories. A food's standard is the description of one or more qualities used to classify it into one of two or more grades in the marketplace. All grading is based on some set of criteria. Standards have been developed for the classification of eggs. The United States Department of Agriculture has set standards for grading these products based on their external and internal quality. In India, egg grading is done in accordance with the Egg Grading and Marketing Rules of 1968. Once eggs and their containers have been graded, they are labelled with the grade and the term Agnark. Taking into account both the shell and the yolk, the Bureau of Indian Standard (BE) has just recently developed standards for grading eggs.

External qualities such as shell and structure, soundness of the shell, and shell cleanliness are taken into account when determining an egg's grade. The typical egg has an oval form with a shell that is sturdy, clean, and devoid of thin patches.

9.2 Exterior Egg Quality Factors

9.2.1. Shell Shape and Texture

1. Practically normal: A shell that is approximately the usual shape and that is sound and free from thin spots. Ridge and rough areas that do not materially affect the shape and strength of the shell are permitted (AA or A quality).
2. Abnormal: A shell that may be somewhat unusual or seriously misshapen, faulty in soundness or strength and showing pronounced ridges and thin spots will be downgraded.

9.2.2. Soundness of Shell

1. Sound: An egg with its shell unbroken
2. Check or crack: An egg that has a broken shell or crack in the shell but with its shell membranes intact and the contents do not leak. Checks may range from a very fine hair like crack (blind or body check) that is discernable only by candling. Eggs with blind checks will not keep well nor can stand even moderate rough handling and so should be used immediately.
3. Smashed: An egg whose shell is smashed or shattered.
4. Leaker: An egg that has a crack with a break in the shell and shell membrane to the extent that the egg contents leak out.

9.2.3. Shell Cleanliness

1. Clean: A shell that is free from foreign material and from stains. An egg may be considered clean if it has only very small specs of stains and dirt.
2. Moderately stained: A shell which is free from prominent stain and adhering dirt, but which has moderate stains.
3. Dirty: A shell which is stained prominently beyond 1/4th of its total surface area. A dirty egg will be downgraded.
4. Adhered dirt: Shell which have foreign dirty material stuck on it like faeces etc. This will be totally downgraded or discarded.

9.3 Candled out Egg Quality

The interior quality of an egg without breaking it can be detected by candling the egg. For this we need a candler. The egg is held gently before the light that emanates from the candler and a gentle twist or twirl is given to the egg to observe the air cell and the yolk shadow. The air cell depth is measured using this candler and a cell depth gauge. The air cell is measured in mm at the broad end of the egg.

9.3.1. Air Cell

1. Practically regular: In this, the air cell maintains a fixed position seen at the broad end of the egg. The outline of the air cell is clear in a white shelled egg.

   Air cell depth: A fresh egg will have an air cell depth of less than 3mm (0.3 cm). As the egg ages, the depth of the air cell increases.
2. Free air cell: An air cell which moves freely between the intact membranes towards the uppermost point of the egg. This will downgrade the egg.
3. Bubbly air cell: A ruptured air cell inner shell membrane resulting in one or more small separate air bubbles usually floating in the albumen (B quality).

9.3.2. Yolk

The yolk outline from egg candling is defined as:

1. Outline indistinct or slightly defined: The yolk outline is indistinctly visible and appears to blend with the surrounding white as the fresh egg is twirled (AA grading).
2. Outline fairly well defined: The yolk outline is visible but not clearly out lined as the egg is twirled (A quality).
3. Outline plainly visible: The yolk outline is clearly visible as a dark shadow when the egg is twirled (B quality).
4. The yolk size and shape determined by grading egg candling is defined as (i) enlarged and flattened - A yolk in which the yolk membranes and tissue have weakened and/ or moisture has been absorbed from the white to such an extent that the yolk appear definitely enlarged and flat (B quality).

The term used to describe yolk defects are

1. Practically free from defects: The yolk shows no blastoderm development but may show other very slight defects on its surface (AA and A quality).

2. Serious defects: The yolk may show well developed germ spots and other serious defects (B quality).
3. Clearly visible germ development: Development of the blastoderm in the yolk that has progressed to the point where it is plainly visible, as a circular spot on the yolk with no blood (B quality).
4. Blood due to germ development: Blood carried by development of the blastoderm in a fertile egg to a point where it is visible as a definite line or a blood ring. These eggs are discarded.

9.3.3. Albumen

The condition of the albumen is determined by candling as follows:

1. Clear: The albumen is free from discolouration of foreign bodies floating in it. Prominent chalazae should not be confused with meat spots (AA quality).
2. Thickness; Fresh egg will have firm thick albumen which keeps the yolk in central position. As egg becomes old, the albumen will become thin and watery and the yolk will be freely moving.
3. Meat spots and blood spots: May be on the surface of the yolk or floating; it is greater than 0.3cm (1/8 inch) diameter (B quality) or if it is greater than 0.3cm (1/8th) diameter, the egg is discarded.
4. Bloody white: Blood has diffused through the white. This egg cannot be consumed hence discarded.

9.4. Grading of Eggs

Table 9.1: AGMARK weight specification

Grade designation	Color Code	Minimum weight of an egg			
		Hen egg		Duck egg	
		OZ	gm	OZ	gm
Extra large (E.L.)	White	2	56.699	$2^{1/2}$	70.874
Large (L)	Red	1.75	49.612	2	56.999
Medium (M)	Blue	1.5	42.524	1.75	49.612
Small (S)	Yellow	1	28.350	1.5	42.524

Source: Agricultural Marketing act - 1937 as ammended in 1986

Table 9.2: Bureau of Indian Standards (BIS grades) specifications for eggs

Grade	Weight/ egg	Shell	Air cell	White	Yolk
A. Extra large	60 and above	Clear unbroken and sound Shape is Normal	Uniform up to 4 mm depth. Perfectly Regular	Clean reasonably Firm	Fairly well centered
A. Large	53-59				Practically free from defects
A. Medium	45-52				-do-
A. Small	38-44				Outline indistinct
B. Extra large	60 and above	Clean moderately stained. Shape is slightly abnormal	8 mm depth May be free from and slightly bubbly	Clear may by slightly weak	May be slightly off centered outline slightly over
B. Large	53-59				
B. Medium	45-52				- do -
B. Small	38-44				- do -

Table 9.3: Summary of U.S. consumer grades for shell egg

US consumer grade	Quality retained	Tolerance %	Permitted quality
Grade AA	87% AA	Up to 13 not over 5	A or B checks
Grade A	87% A or better	Up to 13 not over 5	B checks
Grade B	90% B or better	Not over 1	Checks
Destination Grade AA or fresh fancy	72% or AA	Up to 28 not over 7	A or B checks
Grade A	82% A or better	Up to 18 not over 7	B checks or dirty
Grade B	90% B or better	Not over 10	Checks, dirty, leaker

Table 9.4: Summary of U.S. consumer grades for shell egg

US consumer grade	Quality retained	Tolerance %	Permitted quality
Grade AA	87% AA	Up to 13 not over 5	A or B checks
Grade A	87% A or better	Up to 13 not over 5	B checks
Grade B	90% B or better	Not over 1	Checks
Destination Grade AA or fresh fancy	72% or AA	Up to 28 not over 7	A or B checks
Grade A	82% A or better	Up to 18 not over 7	B checks or dirty
Grade B	90% B or better	Not over 10	Checks, dirty, leaker

Table 9.5: US weight classes for consumer grades for shell eggs

Sizes of weight class	**Egg wt. g/ egg**	**Minimum weight per. Dozen 0Z**	**Minimum net at/ 30 dozen (Ib)**	**Minimum for individual eggs or wt. rate/ dozen 0Z**
Jumbo	70	30	56	29
Extra large	63	27	50 1/2	26
Large	56	24	45	23
Medium	49	21	39 1/2	20
Small	42	18	34	17
Peevee	35	15	28	14

Table 9.6: Summary of United States standards for quality of individual shell eggs specification for each quality factor

Quality factor	**AA quality**	**A quality**	**B quality**	**C quality**
Shell	Clean, unbroken practically normal	Clean, unbroken practically normal	Clean to slightly stained, unbroken may be abnormal in texture and shape	Clean to moderately stained unbroken. May be abnormal
Air cell	1/8 inch or less in depth	3/16 inch depth	3/8 inch or less in depth. May have unlimited movement and be free or bubbly	May be over 3/8 inch in depth. May be free or bubbly
Albumen	Firm, clear	Less firm, may have minute spots	Water with small meat and blood spots	May be weak and watery. May have small blood spots (less than 31 Haugh units)
Yolk	Outlines slightly defined. Practically free from defects	Outline fairly well defined. Practically free from defects	Outline purely visible. May be enlarged and flattened. May show definite defects. Clearly visible germ plasm development but not blood associated with development	Outline plainly visible May be enlarged and flattened. May show clearly visible germ development but no blood. May show other serious defects.

Table 9.7. British standards for shell egg's quality (Showing minimum permissible standards for each class)

Quality factor	Class A	Class B	Class C
White	Normal, clean, undamaged	Dirty, damaged, tinted	Dirty, damaged, tainted
Air space	Normal, clean, undamaged	Normal, damaged	Cracked, misshaped, rough texture on any other abnormality
Air space	Height not exceeding 6mm, stationary	Height not exceeding 9mm, mobile	Height exceeding 9mm damaged
Albumen	Clean, firm or a gelatin like consistency, free on all foreign bodies of any kind	Clear, firm, free of all foreign bodies of any kind	Clean no discontineration or turbid, small foreign bodies permissible
Yolk	Visible on candling as a shadow only without clean out dry hot many appreciably away from the centre of the egg on rotation, free from foreign bodies of any kind	Visible on candling as a shadow only, free from all foreign bodies of any kind, small foreign bodies permissible	Distinct on candling 'siled' or stun, not discoloured
Germ cell	Imperceptible development	Imperceptible development	Imperceptible development
Smell	Free from foreign smell	Free from foreign smell	Free from foreign smell
Wet or dry cleaning	Not permitted	Permitted	Permitted

10

Handling, Processing, Packaging andTransportation of Eggs

Wilfred Ruban S

Abstract

Eggs are naturally Protected by a protective shell, yet they are still quite delicate. Shell breakage might cause significant losses. Specialized packaging and handling processes are needed to ensure the eggs are safe and can be sold for a reasonable price.

10.1 Introduction

The primary purpose of egg packaging is to keep the egg safe from damage, such as cracking, while in transit. By limiting gas exchange via the shell and shell membrane, the egg package may help safeguard the egg's internal quality. Once an egg has been deposited, it is immediately exposed to the conditions of the henhouse. In this environment, it is exposed to many different microorganisms. Almost often, there are no negative effects on health. However, there are microbes that can ruin eggs and others that can cause food poisoning. Salmonella bacteria are found in the digestive tract of all animals and can contaminate the poultry houses. Eggs may be distributed without risk of physical damage or spoiling if they are packaged properly.

10.2 Functions of Packaging

Packaging is an important component in delivering quality eggs to buyers. It embraces both the art and science of preparing products for storage, transport and eventually sale.

Packaging Protects the Eggs from

1. Shell breakage
2. From air cell becoming tremulous or bubbly
3. Micro-organisms, such as bacteria and molds

4. Natural predators
5. Loss of moisture
6. Tainting
7. Temperatures that cause deterioration, and
8. Possible crushing while being handled, stored or transported

The primary purpose of egg packaging is to safeguard eggs against damage, such as cracked shells, during transport and storage. By decreasing the amount of air that may pass through the egg's shell and membrane, the egg's protective packaging may also help preserve the egg's internal quality. Upon being deposited, an egg is immediately subjected to the conditions of the henhouse. It can take up many different kinds of bacteria in this environment. There are very few negative effects on health. But there are microbes that can ruin eggs and others that can cause food poisoning. Salmonella germs, which may be found in the intestines of any animal, pose a threat to chicken farms. In order to minimise physical damage and spoiling during transport, eggs can be properly packaged.

Many factors must be taken into consideration for packaging eggs. It is important to obtain information regarding the necessary requirements for a particular market, such as:

1. Quality maintenance
2. Storage facilities
3. Type of transport
4. Distance to be travelled
5. Climatic conditions
6. Time involved and
7. Costs

10.3 Egg Packages

There are many different types of egg packages, which vary both in design and material.

Type 1: Packing eggs with clean and odourless rice husks, wheat chaff or chopped straw in a firm walled basket or crate greatly decreases the risk of shell damage. The basket has no cushioning material such as straw and therefore damage to the eggs may occur more easily. This kind of packaging may be fit for short distance transport.

Type 2: A very common form of packaging is the filler tray. The fillers are then placed in boxes or cases. Filler trays are made of wood pulp moulded to accommodate the eggs. They are constructed so that they can be stacked one on top of the other and can also be placed in boxes ready for transport. Filler trays also offer a convenient method for counting the eggs in each box, without having to count every single egg. Usually the standard egg tray carries 36 eggs. Therefore, if a box holds five trays, for example, the box has a total of 180 eggs (36 x 5 = 180).

The cases used may be made of wood; however, they are more commonly made of cardboard. When using cardboard cases, special care must be taken in stacking so that excessive weight is not placed on a case at the bottom of a stack.

Fillers can also be made of plastic. The advantages of using plastic egg fillers are that they can be reused and are washable. The fillers can be covered with plastic coverings and be used as packages for final sale to the buyer. More importantly, however, plastic transparent fillers allow for the inspection of eggs without handling or touching the eggs.

Type 3: Eggs can also be packed in packages that are smaller and specific for retail sale. Each package can hold from two to twelve eggs. These cases can be made of paperboard or moulded wood pulp or can be made of plastic.

It is also possible to pack eggs in small paperboard cases and cover them with plastic film. Egg cases have also been developed from polystyrene. The advantages of using polystyrene are superior cushioning and protection against odours and moisture. The package is also resistant to fungus and mould growth.

- The use of small cases is restricted by availability and cost considerations. However, small cases are good for retailers and customers. They are easy for the retailers to handle and customers are able to inspect the eggs.

10.4 Labeling

Labels are a source of important information for the wholesaler, retailer and consumer and not just pieces of paper stuck onto cartons or boxes. The important facts on the label contain information for buyers concerning the eggs, their size and weight and quality/grade description. Labels may also indicate the producer, when the eggs were laid, how to store them and their expiration date. Persuading the buyer to purchase the product without tasting, smelling or touching is another function of labeling.

10.5 Packaging Materials

The modern egg packaging materials are development of egg carton invented in 1911 by newspaper editor Joseph Coyle of British Columbia, to solve a dispute between a local farmers and hotel owners over the farmer's eggs often being delivered broken. The egg carton Box developed by H.G.Bennett, Riseley, UK during the 1950s became popular material for egg packaging.

The following are the common egg packaging and egg transportation materials used now-a-days.

1. **Filler flats (egg trays):** Filler flats are ideal for transportation of eggs. They are stackable and reduce the transport space and cost. They are made up of plastic, recycled paper or moulded pulp. The three sizes available are
   a) Standard – 30 eggs capacity; 292 X 292 cm size.
   b) Compact – 30 eggs capacity; 280 X 280 cm size
   c) Jumbo – 20 eggs capacity; 292 X 292 cm size
   d) Small for quail eggs – 72 eggs capacity; 300 X 300 cm size

   Compact egg trays are suitable for transporting medium chicken eggs and eggs of guinea fowl, native chicken, fancy chicken etc. Standard size egg trays are suitable for large and extra large chicken eggs and duck eggs from Runner and native ducks. Jumbo trays are used for jumbo chicken eggs, Pekin and other meat duck egg and turkey eggs. Small trays for quail eggs are made up of paper pulp or polystyrene or plastic. The paper quail egg trays are wrapped by bubble wraps before stacked in carton boxes. However, plastic trays are stackable therefore no secondary packaging materials like bubble wraps are not needed in bulk carton packing.

2. **Egg shipping cartons:** Carton boxes are the corrugated cardboard egg shipping boxes designed to hold single stack of 12 trays or double stack of 24 trays. Single stack carton box that can hold 12 trays measures 12" X 12" X 13.5" and can accommodate 15 dozen (180) eggs or 10 dozen of jumbo / turkey/ large duck eggs. Double stack cartons measure 12" X 24" X 13.5" and can hold 30 dozen (360) eggs or 20 dozen (240) jumbo/ turkey/ large duck eggs. They come as pre-printed with egg board compliant printing.

3. **Egg cases:** Egg cases are made of strong durable plastics which can last a lifetime. They are very handy in transporting eggs to local market from the farm by non-sophisticated transport. These cases nest easily when not in use. This unit holds up to 15 or 30 dozen Eggs. This unit can be used with any 12" x 12" egg tray as well as standard 12-egg foam and recycled pulp egg cartons. Nestable for easy shipping or storage. They come with different colors for easy identification.

4. **Egg tray sleeves:** These egg sleeves hold 2 1/2 dozen eggs. For use with the regular 5 x 6 egg trays. One egg tray slides inside to make a 30 egg pack.
5. **Egg cartons:** An egg carton is a carton designed for carrying and transporting whole eggs. These cartons have a dimpled form in which each dimple accommodates an individual egg and isolates that egg from eggs in adjacent dimples. This structure helps to protect eggs against stress exerted during transportation and storage by absorbing shock and limiting the incidents of fracture to the fragile egg shells. An egg carton can be made of various materials, including foamed plastics such as Styrofoam, clear plastic or may be manufactured from recycled paper and moulded pulp.
6. **Shipping cases for quail egg trays:** These shipping boxes are 12" x 12"in size. The height varies from 4" to 12" depending upon the number quail egg trays to be packed.

10.6 Egg Baskets

Egg Baskets are made with materials like wire mesh and bamboo and plastic. Wire mesh egg baskets are used for the purpose of egg washing also. Baskets made with coated wire mesh cushions the eggs to protect them from breakage and protects basket from corrosion when washing eggs. They are provided with a handle and available in attractive colours. Egg baskets are inexpensive and the tapering design allows nesting of multiple baskets. The egg basket meant for chicken eggs may be 8", 10" or 12" wide across the top, with the capacity to hold 3, 8 and 15 dozens of eggs respectively. Quail egg basket has a wire small size mesh in the bottom so as to hold quail eggs securely in place. A quail egg basket measuring 10" across the top of the basket, 8" across the bottom and 6" high can hold 12-15 dozen quail eggs.

10.7. Transportation of Eggs

For the successful transport of shell eggs three essential requirements must be met.

1. The containers and packaging materials must be such that the eggs are well protected against mechanical damage.
2. Care should be taken at all stages of handling and transport. Workers handling eggs should be instructed so that they appreciate the need for careful handling. The provision of convenient loading platforms at packing stations, loading depots and railing stations, and handling aids, such as hand trucks and lifts, are of great help.

3. The eggs must be protected at all times against exposure to temperatures that cause deterioration in quality as well as contamination, especially tainting.

The permissible range of temperatures during loading and transport depends on the local climatic conditions and the duration of the journey. Care is needed to avoid excessive shaking, especially where roads are bad. Egg containers should be stacked tightly and tied down securely to minimize movement. Covers should be used to protect them from the heat of the sun, rain and extreme cold where applicable. Where bicycles are used, a device such as a special carrier suspended on springs may be helpful.

A basic prerequisite for all long-distance transport is that arrangements be made for proper reception, handling and storage at the end of the journey. This is especially important where large lots are delivered to a relatively small market. Without access to suitable storage facilities, the eggs may have to be marketed quickly under adverse climatic conditions, which may cause substantial quality deterioration and price losses. Delivery of high quality eggs over long distances, especially in hot climates, generally calls for refrigeration. Requirements for the successful operation of refrigerated transport equipment are rather rigid especially as regards the following factors:

1. Efficiency and durability of insulation;
2. Adequacy and reliability of the cooling mechanism; and
3. Adequate circulation of air within the vehicle or container so that variations of temperature are slight.

Decisions on the establishment of new refrigerated transport services for eggs should be based on thorough economic as well as technical evaluations. The following criteria should be taken into consideration.

1. The need for a managerial and operational staff that is competent in all the operations involved in assembling, loading and distributing.
2. The necessity of a sufficient volume of trade throughout the year.
3. The possibility of making up loads with other compatible produce, e.g. dairy products.
4. The possibility of carrying return loads, once eggs have been distributed.
5. The degree to which the demand for refrigerated transport is concentrated geographically.

10.8. Modes of Transportation

Road Transport: Eggs are transported by road through lorries. For example, a 10 wheel lorry can transport about 2,50,000 eggs. Refrigerated container vans can hold between 3,50,000 and 4,50,000 eggs. These vans are specifically constructed/designed to transport fragile items like eggs.

Rail Transport: Unlike Chicken eggs, duck eggs are normally having thicker shell membranes. Hence it becomes easier and comfortable to transport these duck eggs to various states like Kerala and West Bengal in Brake Vans.

By Sea: Eggs are transported in heavy containers where the eggs are placed in small rectangular boxes totaling to an average of 1500 to 1600 boxes, carrying 3,50,000 to 4,50,000 eggs in each containers.

By Air: Hatching eggs are properly packed and transported.

11

Egg Washing

Kiran M and Wilfred Ruban S

Abstract

In the poultry business, the acceptability of egg washing is contested. It has been established that egg washing can lower the amount of microorganisms on the egg's shell; yet, under some conditions, it can transfer food-poisoning and spoiling organisms from the shell into the egg's inside. The European Union restricts the washing of eggs, although the United States and Japan have no such regulations. The growth of egg-associated food illness, along with an increase in egg production systems' litter systems, has spurred technical interest in methods to enhance the microbiological quality of eggs.

11.1 Introduction

Egg contents may be contaminated via vertical or horizontal paths. In the former, bacteria may colonise the ovaries and/or oviduct, whereas in the latter, pollutants, such as faecal matter or dust, travel through the shell and colonise the membranes and yolk. The surface of a nest-fresh egg provides minimal protection against bacterial pollutants, resulting in the dominance of Gram-positive Micrococci in the shell microflora. Nonetheless, a variety of conditions have been demonstrated to either increase the survival of bacteria in the unfriendly environment of the egg's shell surface or to impair the egg's antimicrobial defence mechanism in other ways. These include the physical state of the cuticle and underlying shell, the presence of water on the shell, the quantity of iron in the water that contacts the egg, and contamination of the shell with organic material such as faeces.

11.2 Egg Contamination

11.2.1 Water on the Shell

The presence of water is crucial when contemplating the contamination of egg contents by faeces and other similar pollutants, and it is especially crucial when considering egg washing. The organic covering, or cuticle, that covers the egg's shell is its primary defence against water and related suspended pollutants. The cuticle permits the diffusion of gaseous water through the shell while preventing the diffusion of liquid water. When viewing the egg soon following oviposition, the significance of water as a factor that facilitates bacterial contamination of eggs becomes very apparent. At this stage, the cuticle is wet and the egg's temperature is normally around 42 degrees Celsius, making it warmer than the surrounding environment. The interior contents constrict as a result of the egg's chilling, resulting in a negative pressure inside the egg. This pressure differential is equalised by drawing air (or water) from the shell's outside into the egg's inside.

11.2.2 Fresh Water Iron

The presence of iron in the wash water was shown to be connected with an increase in egg spoiling. Subsequent studies demonstrated that injecting Fe3+ into the albumen or adding iron salts to the membrane with a bacterial inoculum would result in the unrestricted growth of bacteria, as the iron chelating properties of the ovotransferrin would become saturated, thereby preventing a bacteriostatic effect.

11.2.3 Feacal Contamination

It has been found that there is only a weak association between the amounts of bacteria on the shell and the degree of soiling that is present on the egg's surface. A lot of research have come to the conclusion that there is no direct correlation between the number of pathogens found in faeces and the number of pathogens found in the contents of eggs. However, only a few investigations have provided conclusive evidence that Salmonella enteritidis was the source of the contamination of eggs.

11.3 Common Bacterias in Eggs

Gram-positive cocci and bacillus, such Micrococcus and Arthrobacter, tend to predominate as the most common pollutants on the shell. Gram-positive organisms make up the majority of the contaminants that are found on the shell of eggs. On the other hand, the most of the contaminants that are found inside the eggs are Gram-negative species such Alcaligenes, Achromobacter, Pseudomonas, and Escherichia. It is possible for *S. enteritidis* to infect both the oviduct and the ovaries, which would allow for vertical transmission. However, faecal contamination is thought to be the source of other salmonellas that have been identified from egg shells.

11.4 History of Egg Washing

In the European Union (EU), it is forbidden to wash eggs intended for consumption at the table that are labelled as Class A. However, washing eggs intended for use in industries other than food is permissible. However, nations such as Japan, Australia, and the United States of America all permit and even promote the practise of washing table eggs. In the early part of the twentieth century, a series of tests confirmed that wetting or washing eggs before storage could have an adverse effect on the microbiological quality of the egg. As a result, the European Union has been reluctant to allow the washing of eggs since that time. This reluctance may have its roots in those early tests. Recognizing the issues that were connected with egg washing, the British Egg Marketing Board (BEMB), in the early 1960s, barred packers from washing Class A table eggs. This regulation was put into place. Dry washing was permitted under the Clean Egg Scheme that was implemented in Northern Ireland in 1963. However, egg producers were obliged to provide certification that their eggs had not been exposed to any type of wet cleaning.

The USDA report concluded with a list of recommendations for cleaning eggs which, it was felt, would considerably improve the washing of eggs.

1. Do not attempt to clean excessively dirty eggs
2. Avoid the use of wash water containing >2 ppm iron
3. Do not re-circulate the wash water
4. Use odourless cleaning materials
5. Wash eggs as soon as practical after they are laid
6. Maintain wash water at a temperature that is at least 20°F (equivalent to 11°C) higher than that of the eggs through all washing operations (wetting, cleaning and rinsing)
7. Moisten eggs with stained shells and adhering dirt before eggs are submitted to cutting-spray wash and brushes
8. Precede brushing with a water spray with sufficient force to cut away loose dirt
9. Augment abrasive power of ordinary brushes with abrasive material in brush bristles
10. Maintain an accurate control of the sanitizer-detergent level within the wash water
11. Use a final rinse for the washed eggs
12. Dry washed eggs completely before packing them

11.5 Egg Washing

Wetting, washing, rinsing, and drying are the four processes that make up the overall process, which may be broken down as follows: (i) washing; (ii) rinsing;

and (iii) drying. On the conveyor that moves them through the various steps of the cleaning process, the eggs are normally arranged in a predetermined pattern before being placed there. It is helpful to perform some preliminary washing on the eggs before the primary washing procedure in order to facilitate the loosening of material that may be adhered to the egg shell. The primary step in the cleaning procedure consists of giving the eggs a light scrub using rotary or reciprocating brushes while the eggs are being sprayed with water that is between 40 and 50 degrees Celsius (high pressure water jets alone have been used in some instances). The conveyor often includes rollers that revolve the eggs, which exposes all sides to the brushes or water jet as the eggs pass through the washer. This ensures that the eggs are thoroughly cleaned. It is possible for a washer to give the impression of being a single operation from the outside, but on the inside, there may be two or three separate zones or stages of washing with rising water temperatures (for example, 40-42 degrees Celsius). Rinsing the egg in clean, hot water is the last step in the wet section of the procedure. This is done to get rid of any loose material that the egg may have picked up while going through the main washing process. This step also serves the purpose of removing any chemical or other dissolved matter residues that may have been left behind in the wash water. At this point in the procedure, the egg is given a bath in water heated to temperatures of up to 60 degrees Celsius, making use of the additional heat to facilitate the drying process that comes next. There are at least two stages of operation involved in the drying process. It is accomplished by the combination of two unique physical processes: I the removal of 70–80% of the surface water carried by the egg through the use of mechanical mechanisms, and (ii) the removal of the majority of the remaining water through the use of evaporative mechanisms. Because the first step involves an aspect of drainage, which is typically facilitated by the use of air jets, the rotation of the eggs may be briefly halted during this phase of the process in some designs. The very same air jets that boost evaporation also help the process along by dehumidifying and pre-warming the air that is being evaporated.

11.6 Considerations During Egg Washing

The effectiveness of the egg washing procedure depends on how the water is administered (e.g., temperature, length, use of brushing jets), as well as the water's quality, because water can make it easier for germs to get through the egg shell and into the egg's contents. Water temperature, wash water water quality and mineral content, wash chemicals, wash water pH, and the usage of brushes and jets are the main factors that affect egg washing. In order to avoid the temperature difference creating a negative pressure in the egg that would attract wash water into the egg, the temperature of the final rinse water should likewise always be a little higher than the temperature of the wash water. Potable water should be used throughout the egg washing process to prevent contamination of the egg with organic or inorganic contaminants. For washing eggs, soft water with less calcium is best since it will slow down the rate of calcification on washing machine parts caused by calcium released by shell fragments. When washing eggs, chemicals should be safe for consumption and compatible with the washing machine, the eggs, and other chemicals. For sanitising eggs, chemicals including chlorine and quaternary ammonium are frequently utilised. Commercial egg washing machines have also been employed with other substances, such peracetic acid. Chemicals that may be useful in lowering the bacterial burden, however, could harm the egg's cuticle or shell.

12

Egg as Functional Food

Sudheer K

Abstract

Most authors define the functional food as food containing active components which exert positive effect on physiological processes in the organism and have beneficial effects on human health because they reduce the risk of incidence of various diseases. Functional food, in addition to its basic nutritional value, must have higher content of certain substances (vitamins, fatty acids, antioxidants, probiotics, prebiotics, etc.) which have positive impact on consumers.' health It must be in the shape and form in which it is usually consumed, in other words it cannot be in a form of pill, capsule or similar and the need/requirement to consume greater quantities of the foodstuff than usual in order for the beneficial effect to be exerted is also not allowed.

12.1 Introduction

The eggs represent very good basis that can be enriched with useful and beneficial substances and become functional food, since certain beneficial substances which layers consume in their daily diets transfer relatively easily into their products. However, enrichment of poultry meat and eggs can have adverse consequences in regard to their quality. Functional food is interesting also from the economical aspect, because it opens for poultry producers the possibility to introduce new products into their product range. Of course, main assumption is that consumers of enriched products are willing to pay premium price for these products, which is by 15-20% higher than average product price. On a global basis, eggs are one of the most recognized and accepted foods by consumers. They are naturally extremely nutritious and widely acknowledged as a great source of protein and selected vitamins and minerals. Numerous studies have demonstrated that the nutritional content of eggs can be easily and repeatedly modified to enrich them in various nutrients, making the eggs one of the most natural functional foods. A number of health benefits derived from the consumption of eggs enriched with omega-3, lutein, vitamin E and selenium have been clearly demonstrated in sound clinical studies.

For example, the consumption of omega-3 enriched eggs has been shown to have some positive effects during pregnancy and lactation of infants. Eggs with additional omega-3 were also shown to benefit patients already under medication to control high cholesterol levels. Finally, the ease and natural way to produce functional eggs are further strengths for their use in human nutrition. Unfortunately, while functional foods are more readily accepted when the base product is considered healthy, the attitudes of consumers towards eggs and especially on the issue of cholesterol still limit the use of functional eggs. Surveys in a number of countries demonstrate that consumers are extremely ambivalent about eggs, consuming more eggs and fewer eggs for the same health reasons. Other issues like the lack of visual differentiation of functional eggs, the consumer disbelief about information provided by marketers and the lack of enthusiasm by the medical profession to recommend the consumption of functional eggs are slowing the beneficial impacts that functional eggs can make for human health. However, with our aging population and the rising cost of health care, it is fair to consider that functional eggs can play a bigger role for human health. In addition, nutritional gaps like that created by the low consumption of fish in several Western societies create some opportunities for functional eggs enriched with omega-3. Regardless, consumer education along with that of health professionals will likely be one of the keys to the success of functional eggs for human health.

12.2 Eggs as an Integral Part of the Diet

Hens' eggs have been used as a food by human beings since antiquity. Compared with the egg, no other single food of animal origin is eaten by so many people all over the world and none is served in such a variety of ways. Its popularity is justified not only because it is easy to produce and has so many uses in cookery, but also because of its nutritive quality. Of the three dietary essentials proteins, fats, and carbohydrates, the egg is composed largely of the first two. Furthermore, the nutritive quality of an egg enhances the value of any food in which it is incorporated. Its wide use in cookery for purposes of leavening, thickening, binding, and emulsifying considerably improves the human diet. Egg proteins are highly digestible containing the most important essential amino acids in a profile that is not dissimilar to the ideal balance of amino acids needed by men and women. Eggs also supply various minerals, some in significant amounts, and contain major vitamins. Many of those nutrients in the egg can be manipulated by dietary means; however, a real value for improvement of human diet could have only those nutrients which are usually in short supply with other products or have a positive effect on human health when consumed in excess. Among them n-3 fatty acids, vitamin E, carotenoids and selenium have attracted a lot of attention in nutritional sciences.

12.3 Points to be Considered Before Enriching Eggs

- Efficiency of nutrient transfer from feed to the egg.
- Availability of commercial sources of effective feed forms of the nutrient
- Possible toxic effects of nutrients for the laying hens (Ex: Vitamin A and D are toxic for chickens at high levels)
- Amount of nutrient delivered with an egg in comparison with Recommended Dietary Allowance (RDA).
- Established health promoting properties of nutrients and their shortage in modern diet.
- Possible interactions with assimilation of other nutrients from the egg.
- Stability during cooking.

12.4 Dietary Manipulations to Produce Designer Eggs

Dietary manipulation is the major step in producing the preslaughter value added poultry products. Along with the regular feed ingredients required for production of eggs certain compounds which induce certain functional benefits in the eggs may be incorporated in the diet.

12.4.1 Omega-3 Fatty Acids Enrichment

The total fat content in the egg yolk cannot be altered but its fatty acid composition can be altered, by changing the type of oil used in the hens' diet. Omega-3 fatty acids are a group of polyunsaturated fatty acids (PUFA) also known as n-3 fatty acids. Majority of birds and animals including human beings cannot synthesize these n-3 fatty acids, hence they are dietary essential. Sources of Omega-3 fatty acids are fish oil, linseed/ flaxseed & oils, rapeseed / canola & oils, safflower oil, soyabean oil and marine algae. Unfortunately, the foodstuffs rich in these n-3 fatty acids namely linseed and fish oils have undesirable flavor and consistency, hence not relished commonly by human beings. They can be made acceptable by incorporating these beneficial n-3 fatty acids into eggs. Commercial table eggs contain a high proportion of n-6 PUFA (mainly 18:2n-6) but are a poor source of n-3 fatty acids. Attempts to produce eggs high in n-3 PUFA can be divided into two groups. The simplest way is to produce an egg enriched in linolenic acid, which is a precursor of DHA and is also considered to have a protective effect against fatal ischemic heart disease. For this purpose, the hen's diet is usually relatively rich in flaxseeds, linseeds or their corresponding oils; as a result, the egg's yolk is enriched with alpha linolenic acid (ALA) and the level of DHA is also enhanced. However, given that most of the health promoting properties of n-3 fatty acids are associated with DHA, the health benefits of ALA-enriched eggs could be limited (since the conversion of linolenic acid into DHA in human body is not always effective). This is especially so in the elderly and children when their diets are rich in n-6 PUFAs. The second group or route to enhancing levels of n-3 in the egg, by including pre-formed DHA in the hen's diet, usually in the form of fish (menhaden, herring or tuna) oil, is a more promising one. However, this may be associated with a pronounced fishy taste in the egg yolk.

12.4.2 Organoleptic Quality of Omega-3-enriched Eggs

The organoleptic quality of omega-3 eggs tends to be similar to regular table eggs although in some cases panellists are able to detect off-flavours. A'fishy' or fish-product related flavour was detected in eggs from hens on diet containing 15-20% flax seed and a study conducted by some researchers suggest that high (>10%) levels of flax seed will result in some decrease in overall egg acceptability as assessed by aroma and flavour. This remains a problem associated with the commercial production of this type of product. It has been suggested that the use of combinations of anti-oxidants in the hen's diet could help to suppress these off-flavours; however, high levels of vitamin E did not prevent the formation of off-flavours in hens fed on the high (>10%) flax diet .In general, therefore, fishy taints in eggs are not detectable provided that the hens are fed 5% (or less) flaxseed or low levels of a high-quality oil, e.g., 1.5% (or less) menhaden fish oil. In this respect, a dried DHA-enriched marine micro-algal product showed promising results. The fishy taint can result from rancidity of the n-3 enriched diet, therefore the quality of oils used is important in eliminating off-flavour in eggs. While the oil can be protected with anti-oxidants or micro-encapsulated once the oil is oxidized the addition of anti-oxidants will not reverse this situation. It is notable that the acceptability of 'fishy' taints varies from country to country. For example, eggs produced in some countries (Chile), where fish meal as well as fish oil are usual components of the chicken diet, contain quite high level of DHA and in many cases have a fishy taste (personal experience of PFS). Generally, the public in Chile are used to this taste and accept it as normal while, for Europeans, for example, it would be unacceptable. Cooking characteristics of omega-3 eggs, including emulsification capacity, hardiness and springiness of sponge cakes prepared using these eggs, were the same as in ordinary eggs. Data on effects of n-3 enrichment on egg quality during storage are limited and sometimes contradictory. For example, it has also been shown that there is no alteration in fatty acid profile of eggs enriched with n-3 PUFAs during cooking or during storage for 7 weeks at 25°C, whereas other studies have reported an increased susceptibility to oxidation during storage and cooking. Enrichment of egg yolk in vitamin E is thought to be an effective means to resolving this problem and has been shown to significantly decrease thiobarbituric acid values (TBARS) in n-3-enriched eggs. In other experiments, yolk TBARS values in n-3-enriched eggs were not significantly different from those in control eggs. However, it needs to be recognized that this parameter depends on the analytical technique used and the level of the egg enrichment with n-3 PUFA. Therefore, attention needs to be paid to the organoleptic properties of n-3-enriched eggs and in particular the relationship between n-3 levels and organoleptic quality.

12.4.3 Clinical Significance of Omega-3 fatty Acids

Research on the various health benefits of dietary omega-3 fatty acids has shown that these fatty acids reduce plasma triglycerides, blood pressure, platelet aggregation, thrombosis and atherosclerosis particularly in diabetics, tumour growth, skin disease, supresses inflammatory process and enhance immunity. In designer eggs the n-6 / n-3 PUFA ratio is decreased to about 1.5 from as much as 20 in regular eggs. Without any change in the sensory quality of the egg this change in designer eggs will supply about 50% of the daily requirement of n-3 PUFA ratio to the consumer. Since n-3 fatty acid enriched eggs will undergo rancidity quickly, it is necessary to prevent the rancidity of the designer egg yolk lipids, by incorporating anti-oxidants in the hens' diet and extend the shelf life of the product.

12.4.4 Anti-oxidants in Eggs

Egg is a rich source of natural antioxidants like vitamin-E, selenium, carotenoid pigments, flavonoid compounds, lecithin and phosvitin. These compounds protect the fat-soluble vitamins and other yolk lipids from oxidative rancidity. However, these levels are not sufficient to protect the designer eggs rich in n-3 PUFA. Hence it is essential to increase the anti-oxidant levels in the designer eggs. The designer egg and meat, not only contain high levels of the above antioxidants; but also rich in herbal active principles like, Allicin, Betaine Eugenol, Lumiflavin, Lutein, Sulforaphane, Taurine Eugenol, Lumichrome, Lycopene, Curcumin, Carnosine, Quercetin, depending upon antioxidants of herbal origin used in hens' diet. Supplementation of these anti-oxidants in hens' diet will increase their levels in the egg and meat.

The advantages of enrichment of the egg and meat with anti-oxidants include:

1. Decreased susceptibility to lipid peroxidation
2. Prevention of fishy taint to the product
3. Designer foods could be a good source of antioxidants in human diet
4. Prevents destruction of fat-soluble vitamins and natural fat-soluble pigments

For designer egg/meat production, vitamin E and organic selenium are added as anti-oxidants at levels of 200-400 mg/kg and 0.1-0.3 ppm respectively. Besides these, other anti-oxidants such as chemicals and herbs may be added, to prevent oxidative rancidity.

12.4.5 Lowering Cholesterol Content

A large egg contains about 200 mg of cholesterol and chicken meat contains about 60 mg per 100 g. Research towards lowering egg cholesterol has been concentrated mostly on dietary and pharmacological interventions. Chromium, copper, nicotinic acid, statins, garlic, basil (tulasi), plant sterols, n-3 PUFA supplementation to chicken feed will reduce the yolk and carcass cholesterol levels significantly.

12.4.6 Enrichment with Carotenoids

Carotenoids occur naturally in egg yolk in varied amounts depending on hen's feed. Deep yellow or orange colour yolks and yellow skin broilers are preferred over pale yolks and skin. Feed fortification with natural sources such as Marigold, Alfalfa extracts are sources of lutein. Other sources such as Corn and Red pepper provide zeaxanthin and capsanthin respectively. Lycopene is a hydrocarbon carotenoid reported to have strong antioxidant properties effective in reducing the risk of prostate carcinoma. Although lycopene is not usually found in eggs, lycopene enrichment can be achieved via feed fortification with tomato powder and lycopene could reduce yolk lipid peroxidation. Besides providing attractive colour, they act as anti-oxidants and anti-carcinogenic agents. Some of the pigments have vitamin A activity. For example, the lutein which safeguards the retina.

12.4.7 Enrichment with Immunomodulators

Chicken egg is abundant in antibodies like IgY which is cheaper and better than mammalian immunoglobulin IgG. This IgY can be used to treat human rotavirus, *E.coli*, *Streptococcus*, *Pseudomonas*, *Staphylococcus* and *Salmonella infections*. The IgY level in the egg can be increased by dietary manipulations. The functional feed rich in omega - 3 fatty acids and anti-oxidants itself will increase the IgY level in the egg. Herbal supplementation will further boost the IgY level in the egg. Among the herbs, Basil leaves (Tulasi) is having the highest ability to boost the IgY level in the egg. Other herbs like Rosemary, Turmeric, Garlic, Fenugreek, Spirulina, Aswagantha, Arogyapacha etc also possess immuno modulating properties.

12.5 Conclusion

Poultry eggs are a good source of essential nutrients. The egg is considered as nature's most complete food containing high quality proteins, a 2:1 ratio of unsaturated fats to saturated fat, an excellent source of iron, phosphorus and other minerals and all vitamins with the exception of vitamin C. Although the best source of all the vital nutrients, high cholesterol content is the major constraint for egg consumers. The development of nutrient enriched value-added poultry eggs greatly increased the context of functional foods for human health. By manipulating the diet of chicken with nutritional interventions like supplementation of chromium, copper, herbs, probiotics, n-3 fatty acids, antioxidants etc. value added and health promoting chicken egg, meat and their products can be made available to the consumers. The designing must take into consideration the production facilities, available materials, technical know-how, economic resources of the producers and environmental impacts with welfare issues.

13

Microbial Challenges and Issues of Egg

Deepika Jamadar and Kiran M

Abstract

The aim of this chapter is to highlight the types of microbial challenges and current issues related to eggs. Under healthy breeding conditions, an egg's contents are generally sterile just after laying. However, they can be contaminated by a diversified microbiota containing food spoilage microorganisms and sometimes pathogenic bacteria. The risk of contamination by Salmonella is a major concern for the egg production and egg product manufacturing industries. Eggs can be contaminated both externally, on the eggshell, and internally, i.e. during development. By decreasing the rate of contamination of the environment, the risk of horizontal and vertical contamination is reduced. Regarding eggs, not only the bactericidal activity of egg white but also the strength of the eggshell have a genetic component. These parameters could therefore be improved by a specific breeding programme, although more information on the antibacterial compounds involved in protection is essential to achieve this aim.

13.1 Introduction

Many changes have occurred in egg production and processing, as well as in the egg itself, over the last 40 years. All of these changes have contributed to the microbial challenges of the egg that we face today. In the past, production traits including egg size, feed conversion and egg output have received the majority of attention in genetic selection of laying hens. Microbial integrity hasn't been given much attention. Until the 1990s, food safety concerns associated with shell eggs were not considered. Eggs were thought to be pathogen free until they were cracked.

Contamination of eggs can occur either by shell surface contamination or by contaminated contents in vivo. *Salmonella enterica serovar Enteritidis* (SE) and more recently *Salmonella enterica serovar Heidelberg* (SH) are the two organisms of most concern associated with eggs.

Under healthy breeding conditions, an egg's contents are generally sterile just after laying. However, they can be contaminated by a diversified microbiota containing food spoilage microorganisms and sometimes pathogenic bacteria. Eggs can be contaminated both externally, on the eggshell, and internally, i.e. during development. The egg may therefore be a vector of bacteria causing foodborne illness in humans: this is particularly the case for Salmonella.

13.2 Microbial Spoilage

The most predominate spoilage of shell eggs is caused by Gram-negative motile rods: *Proteus, Aeromonas, Pseudomonas, Acinetobacter, Alcaligenes, Bacillus, Coliforms, Arthrobacter, Escherichia coli, Serratia, Flavobacterium, Cytophaga, Enterobacter cloacae, Micrococcus, Hafnia*, and *Citrobacter* are additional spoilage microbes that can contaminate eggs. Eggshells commonly associate with the following serotypes of Salmonella: *Salmonella enterica subsp. enterica ser. Anatum, Salmonella enterica subsp. enterica ser. Bareilly, S. Enteritidis, Salmonella enterica subsp. enterica ser. Derby, Salmonella enterica subsp. enterica ser. Essen, Salmonella enterica subsp. enterica ser. Heideberg, Salmonella enterica subsp. enterica* ser. *Montevideo, and Salmonella enterica subsp. enterica ser. Typhimurium. Penicillium, Alternaria*, and *Mucor* can produce different types of mold rots in eggs. Mold spoilage of egg is generally referred to as spots due to the formation of mycelial growth on the eggs. Yellow, green or bluespot are caused by *Penicillium*; black spots or dark green by *Cladosporium*; pink spots by *Sporotrichum*; and other mold genera causing egg spoilage are Botrytis, *Alternaria* , *Mucor*, and *Thamnidium*.

Molds generally multiply in the air sacs of eggs, since oxygen favors the growth of molds. Off-flavors (mustines) in eggs can be caused by *Pseudomonas graveolens, Achromobacter perolens, Streptomyces, Pseudomonas mucidolens*, and others. *Enterobacter cloacae* causes a hay odor and *Escherichia coli* fish flavors. Bacteria produce H_2S and other foul-smelling compounds. Molds growing in the shell also give musty odors and tastes. The predominant bacteria causing spoilage in pasteurized egg products are psychrotrophic Gram-negative. Dried eggs are not susceptible to microbial spoilage due to low water activity (aw).

Table.13.1 -List of microbial spoilage causing organisms.

Organism	Change	Type of rot
Proteus spp.	Dark brown mealy yolk; dark brown albumen	Black rot type 2
Aeromonas liquefaciens	Gelatinous yolk blackened throughout; gray watery instead of "white"	Black rot type 1
Enterobacter spp.	Yolk encrusted with custard-like material and occasionally flecked with olive-green pigment	Custard rot

Serratia marcescens spp.	White-stained red throughout; yolk surrounded with custard-like material	Red rot
Pseudomonas maltophilia	Gelatinous, amber-like yolk striped olive green, almond-like odor	Green rot
Pseudomonas fluorescens	Fluorescent green changing to pink white; yolk surrounded with custard-like material	Pink rot
Pseudomonas putida	Fluorescent green pigment throughout the albumen	Fluorescent green rot
Pseudomonas aeruginosa	Fluorescent blue pigment throughout the albumen -	Fluorescent blue rot Yellow rot
Flavobacterium cytophaga	Yellow pigment formed in membrane at site of microbial growth	Fluorescent blue rot Yellow rot
Other *Enterobacter and Alcaligenes* spp.	Although large populations develop, there are no microscopic changes of the yolk and albumen	Colorless rot

13.3 Microbiological Standards

In addition to any national requirements, the microbiological condition of the egg and egg products shall be in conformity with the following minimum requirements:

Salmonellae

a) Salmonella organisms should not be recovered from any of ten sample units examined when the test is carried out according to the method described (n = 10, c = 0, m = 0)[6].

   n = The number of sample units to be examined. m = The value at or below which no concern is recognized.

b) In products intended for special dietary purposes, salmonella organisms should not be recovered from any of thirty sample units examined (n = 30, c = 0, m = 0)[6].

Mesophilic aerobic bacteria should not be recovered from any of five sample units examined when the test is carried out according to the method described in a number exceeding one million per gram, nor in a number exceeding 50,000 per gram from three or more of the five sample units examined. (n = 5, c = 2, m = 5×10^4 , M = 10^6)[7].

Coliform bacteria should not be recovered from any of the five sample units examined, when the test is carried out according to the method described, in a number exceeding 1,000 per gram, not in a number exceeding ten per gram from three or more of the five sample units examined. (n = 5, c = 2, m = 10, M = 10^3)[7].

13.4 Microbial Challenges

13.4.1 Endogenic Contamination

When eggs are developing in the oviduct, they may become contaminated. Vertical contamination is the term used to refer to this sort of contamination, which is primarily caused by Salmonella. Salmonella contamination can occur throughout the egg-laying process in the ovary or oviduct of infected hens. *Salmonella enteritidis* inoculation of poultry via oral, intravenous, or peritoneal routes has already been attributed by many authors to egg contamination. Although these species have been found after laying eggs, endogenic contamination by *Salmonella typhimurium, Salmonella infantis*, and *Salmonella hadar*, has never been noted (exogenic contamination). Several studies on experimentally infected hens have shown that endogenic contamination may occur at each step during egg development. vertical contamination of eggs by the type I avian Influenza virus, which causes an infection in several species of birds including hens.

13.4.2 Exogenic Contamination

Exogenic contamination of eggs is also known as horizontal contamination (contamination of eggshell surface), which, in terms of overall levels and bacterial variety, occurs far more frequently than vertical contamination. It occurs after laying, when the eggshells are in contact with hen's faecal microorganisms or with other microorganisms present in the farm environment, or elsewhere in the supply chain. Contamination may also happen while eggs are being transported, packaged, or handled on the farm or in the egg packaging site. The environment or another egg could be the source of the infection. Contamination is more likely when there are cracked or 'flowing' eggs present. Controlling the elements affecting eggshell contamination is fundamental to prevent the spread of microorganisms for all the reasons mentioned above.

13.4.3 Penetration of Microorganisms Through the Eggshell

Even if the egg surface is contaminated, the cuticle, shell, and shell membranes act as barriers to stop bacteria from getting inside the egg and contaminating its contents. Mucoproteins make up the film that makes up the cuticle. It is resistant to water and microbial infiltration. However, due to its limited effectiveness of a protective coating, it cracks rapidly during egg handling.

Due to the possibility of microbes passing through its pores, the shell is inefficient to act as a good physical barrier. Though, it includes lysozyme and ovotransferrin, which may help to prevent penetration of microorganisms. The degree of eggshell surface contamination, eggshell cracks, and irregular eggshell calcification are the key characteristics that do seem to influence microbe penetration.

13.4.4 Bacterial Survival and Multiplication in Egg White

Role of antimicrobial molecules: Lysozyme: present in many biological fluids, is responsible for different types of bactericidal activity. Ovotransferrin, it's a protein that helps in chelating metal ions, which has a bacteriostatic effect through the creation of an iron-deficient environment. The significance of egg white's physicochemical characteristics and the presence of other proteins such as protease inhibitors present in egg white, including cystatin (a cysteine protease inhibitor), ovomucoid and ovoinhibitor (serine protease inhibitors), and ovostatine (inhibitor of many proteases). It has also been demonstrated that the vitamin chelating proteins i.ethiamine binding protein (which fixes thiamine), avidin (which fixes biotin), and flavoprotein (which fixes riboflavin) are involved in the antibacterial activity of egg white, although their contribution seems to be minimal.

Influence of environmental conditions: Bacterial growth in egg white and migration to the yolk depends not only on the microbes but also on the storage time, temperature and quality of the eggs.

13.4.5 Bacterial Migration and Multiplication in Egg Yolk

The chalazae and egg white will liquefy over time, and the yolk will no longer be held in the center. As the gap between the egg yolk and eggshell is shortened, facilitating access of microorganisms including bacteria from the eggshell to the yolk reduces, hence promoting the access of microorganisms (including bacteria derived from the eggshell) to the egg yolk. The vitelline membrane, which contains certain proteins such as ovomucin and lysozyme, represents the final obstacle to the migration of microorganisms to the yolk.

13.5 Strategies to Combat Egg Contamination

Various ways to reduce the level of egg contamination have been explored. These include both upstream methods such as hen selection, breeding practices, farm management and downstream methods are packaging, transport and storage of eggs.

13.5.1 Genetic Selection

There is a heritable genetic basis for hens resistance to *S. enteritidis* caecal colonization. Similarly notable variations in the expression of various genes encoding proteins involved in the defense against Salmonella colonization in two lineages of hens. Thus, hen selection may be an efficient way to improve resistance to colonization by Salmonella. Regarding eggs, not only the bactericidal activity of egg white but also the strength of the eggshell have a genetic component. These parameters could therefore be improved by a specific breeding programme, although more information on the antibacterial compounds involved in protection is essential to achieve this aim.

13.5.2 Husbandry Methods

It is well known that birds under stress on farms are more vulnerable to acquiring Salmonella. Additionally, stress plays a role in the excessive calcification of eggshells. Stress tends to weaken the immune system and lower the inflow of macrophages to the reproductive organs during the peak of egg production, encouraging infection of the ovaries.

13.5.3 Breeding Practices

It seems that the size of the flock is the predominant risk factor for hen contamination by *S. enteritidis.* The number of cracked or broken eggs also seems to relate to the type of hen housing system and the design of the cages.

13.5.4 Cleaning & disinfection Practices of the Breeding Environment

Since the level of the mesophilic microbiota present on the surface of eggshells is related to the level of air contamination in the breeding environment, it seems crucial to limit the dust and to promote good hygiene practices. Building decontamination between two hen batches is now carried out routinely.

13.5.5 Hen Vaccination against *S. Enteritidis*

Hen vaccination has been obligatory since 1 January 2008 in countries of the European Union in which Salmonella prevalence in laying hens exceeds 10%. This practice is probably an important preventive way of reducing the level of hen contamination but scientific evidence to verify its effectiveness in reducing egg contamination is lacking.

13.5.6 The use of Feed Additives

A wide range of anti-Salmonella food additives are available. Short chain organic acids (including propionic, butyric and formic acids) or medium chain organic acids (caproic, caprylic, capric and lauric acids) are often used and are intended to reduce colonization of the hen digestive tract and, consequently, faecal excretion of Salmonella. An alternative strategy is the addition of pre or probiotics to the hen diet. Fructo-oligosaccharides are used to promote the growth *of Bifidobacteria*, which may reduce colonization by *Salmonella.* Probiotics are living microorganisms which, when they are administered in adequate amounts, confer a health benefit on the host. Some studies, especially those with lactic acid bacteria, have already been carried out. However, the effects of these additives on levels of egg contamination are not yet clear.

13.5.7 Practices of Egg Collection, Sorting and Storage

To reduce the risk of contamination, good practices must be followed for egg collecting, sorting, packing, storage, and delivery. Cross-contamination must be avoided by monitoring staff and environmental contamination, such as that caused by pests and surfaces. At each step, it is essential to preserve eggshell integrity.

It is necessary to limit temperature changes during egg collection on the farm and delivery to the egg packaging centre. In France, eggs must be stored between 5 and 25 °C. The use of air-conditioned storage rooms and the practice of isothermal transport would minimize temperature variation. It could be interesting to investigate the effects of cooling eggs quickly after laying and maintaining the cold chain until delivery, in order to limit penetration, migration and proliferation of Salmonella in the egg content.

13.5.8 Practices of egg Decontamination

Egg washing is thought to be the cause of the weakening of the egg's surface barriers, such as the cuticle, as well as an increase in humidity that encourages the entry of microorganisms and elevates the level of internal Egg contamination. Studies on egg pasteurisation have been conducted, according to Braden (2006), heat treatment may help to lower salmonellosis cases related to egg consumption in the US.

Other methods of egg decontamination including 'flash' (hot water or steam) treatments, ozone, ultrasound, radiation, UV, ionized air (plasma) or electrolysed water treatments. It is necessity to study the benefits of adopting the method in terms of combating microbiological challenges and emerging issues concerning egg and egg products should be underlined. All egg decontamination processes must be validated under natural contamination conditions at the industrial scale.

14

Designer Eggs

Wilfred Ruban S and Kiran M

Abstract

Designer foods with an animal origin are made either by feeding certain diets to animals or by employing cutting-edge methods like cross-breeding and genetic editing. Among the many types of functional foods, designer eggs are highly valued. Interest in using poultry biotechnology to modify egg composition through genetic and nutritional alterations to benefit human health has increased in tandem with the development of the chicken industry. Incorporating therapeutic pharmaceutical compounds or adjusting dietary factors such as cholesterol and lipid concentration and fractions, lipid profile, fatty acids, amino acids, and minerals, can achieve this goal.

14.1 Introduction

Eggs are an essential food for humans due to the high nutritional quality and quantity that they contain. Eggs are one of the most consumed foods all over the world because of its nutritious profile, range of uses, and relatively inexpensive cost as a food source. Intakes of total lipids, cholesterol, and saturated fatty acids in human diets can be attributed, in part, to the consumption of animal products. Due to the high amounts of cholesterol (about 200-300 mg/100 g) and saturated fat (approximately 3 g/100 g), eggs have been considered to be contentious foods by several health organisations and nutritional experts. Because of these factors, the general public was previously cautioned against increasing their consumption of eggs because of the amount of cholesterol they consume and the possible link between eggs and cardiovascular disorders.

More than 600 milligrammes of omega-3 polyunsaturated fatty acids and 6 milligrammes of tocopherol may be found in designer eggs. Consumers of eggs all across the world may soon have access to an additional option of food product thanks to the functioning egg. Eggs with unique designs are sold in marketplaces all over the world under a variety of titles, each of which corresponds to a certain region. The egg business has done a good job of responding to the unfavourable attitudes of customers regarding the difficulties associated with eggs and their products, notably the cholesterol content, by searching for innovative strategies to alter these perceptions. Retail markets for functional eggs already exist, particularly for eggs that have been fortified with omega-3 PUFA (poly unsaturated fatty acids) or those have a reduced cholesterol level. The majority of the time, these eggs are created by altering the diets of the hens that lay them, but the creation of designer eggs through the use of technical processes has received a lot less attention.

14.2 Production of Designer Eggs

Egg proteins are very easily digested and include all of the main essential amino acids in a profile that is very similar to the optimal balance of amino acids that consumers require. Egg proteins also contain a variety of minerals and vitamins in addition to the required amino acids. According to the findings of a research, one egg contains between 10 and 20 percent of the recommended daily intake (RDI) for total fatty acids, saturated fatty acids, and polyunsaturated fatty acids. Eggs also contain between 20 and 30 percent of the RDI for vitamins E, A, and B12. In this regard, an egg can deliver 10% of the recommended daily intake (RDI) for humans of protein, 6% of the RDI for vitamin A, 6% of the RDI for vitamin D, 3% of the RDI for vitamin E, 15% of the RDI for vitamin B2, 4% of the RDI for vitamin B6, 8% of the RDI for vitamin B12, 6% of the RDI for folic acid, 2% of the RDI for thiamine Changing the nutritional make-up of diets has the potential to boost levels of a significant number of key elements found in eggs, including vitamin E, omega-3 fatty acids, selenium, and carotenoids. Eggs and other chicken products can have their composition and nutritional quality influenced by the nutrient density and balance of the meals of the birds who laid them. The development of functional meals involves the careful application of diet formulation in poultry, and the nutritive contents of components must be sufficient for there to be an influence on public health and welfare as well as to guarantee that excellent quality chicken products are produced. Low-fat meals are now widely available thanks to the efforts of nutritionists, researchers, and food technologists. These efforts are motivated by the broad implications of reducing the risk of certain diseases, such as stroke, cancer, and cardiovascular diseases.

14.3 Importance of Designer Eggs

In addition to contributing to platelet inhibition, anti-ACEI activity, and anti-ROS activity, a normal egg supplies 70 kilocalories of energy in the form of proteins, carbs, fatty acids, amino acids, lipids, cholesterol, and polyunsaturated fatty acids. Eggs also include potassium and salt. Eggs include anti-inflammatory and anti-coagulation properties, as well as anti-hypertensive peptides such as ovokinin and 12 essential minerals and vitamins, all of which work together to help protect organs such as the heart from a variety of disorders. Eggs are an inexpensive source of high-quality protein and also include a balanced ratio of vitamins (with the exception of vitamin C) and minerals; nevertheless, the consumption of eggs on a per capita basis is low since eggs contain a significant amount of cholesterol (200 mg per egg).

14.4 Cholesterol Reduction

Consumers are concerned about the high cholesterol contents of eggs, which average 200 mg/egg on average. This is despite the fact that there are controversies regarding whether or not cholesterol has a negative effect on health and whether or not the consumption of dietary cholesterol can increase blood cholesterol. Several research have been carried out in an effort to lessen the negative effects that are brought about by the presence of triglycerides and total cholesterol in poultry meat and eggs. Dietary manipulations, the use of drugs or natural feed additives, and genetic manipulations have all generally resulted in a reduction of egg cholesterol of no more than ten percent. Although these methods can help reduce egg cholesterol, they have not been very successful because nutritional, pharmacological, and genetic manipulations have only produced a ten percent reduction at most. However, the inclusion of PUFA in poultry diets results in a considerable decrease in the amount of cholesterol and total lipids. Based on the findings of a few research, it appears that egg cholesterol levels may be reduced by around 25% utilising various dietary approaches. It is well known that the use of Medicago sativa is effective in decreasing re-absorption of biliary cholesterol in the intestine of birds, which regulates its whole-body content by decreasing the cholesterol level in the blood and yolk. The efficient role of alfalfa meal as a hypocholesterolemic factor in egg can be partially attributed to lipogenesis inhibition in the liver. This is because it is well known that Medicago sativa is effective The benefits of omega-3 fatty acids and antioxidants (vitamin E, spirulina, and organic Se) in diets have been combined to provide this larger cholesterol reduction, which has been linked to these combined effects.

14.5 Enrichment of Vitamin and Minerals

It is feasible to increase the levels of iodine, selenium, and chromium in eggs by dietary modification without influencing the levels of calcium and phosphorus. The majority of minerals are located in the shell, where calcium and phosphorus are present in high amounts. The creation of selenium-enriched eggs is fairly straightforward and may be accomplished by adding 0.4 mg/kg of Se yeast to generate an egg with around 30 mg of selenium, or nearly 50% of the Recommended Dietary Allowance. Designer eggs can include higher levels of the antioxidant vitamins E and A, which help protect against heart disease.

14.6 Modification of Fatty Acid Profile

There are markets for designer eggs with reduced saturated to unsaturated fatty acid ratios. This ratio is frequently adjusted by the addition of canola oil to animal feed. Balanced ratios of PUFA/SAFA (1:1) or omega-6/omega-3 PUFA can be found in designer eggs (1:1). Omega-3 polyunsaturated fatty acids control the immune system and are essential for the development of the neurological system. In addition to their anticancer, anti-inflammatory, and cardioprotective properties, they reduce blood platelet aggregation, the incidence of thrombosis, blood pressure, and atherosclerosis. By supplementing the diets of laying hens with particular nutritional supplements, such as groundnut oil, fish oil, safflower oil, linseed, fish meal, or algae, the egg omega-3 content can be increased. EPA, DPA, DHA, and LNA are examples of omega-3 fatty acids, whereas arachidonic acid (AA) and linoleic acid (LA) are examples of omega-6 fatty acids. Omega-3 fatty acid consumption through designer eggs for humans can protect against cardiovascular issues, such as heart attack, and can substitute fish products in the diet of the consumer. It has been suggested that high dietary intakes of n-6 PUFA are a major limiting factor in the conversion of C18:3n-3 to EPA and DHA. Standard eggs have a high ratio of n6 to n3 fatty acids.

14.7 Egg Antibodies

Eggs might be used as a source of immunogens that strengthen the immune system of humans, especially children, against a variety of illnesses. It is possible to utilise transgenic chickens to manufacture antibodies in the albumin of their eggs. Egg yolk antibodies (EYA) are antigen-specific antibodies in the egg yolk that may be acquired by injecting hens subcutaneously or intramuscularly with a particular antibody that induces the formation of IgY antibodies. Innovative biotechnology helps the process of creating genetically modified chickens that might be used to produce eggs containing insulin for the treatment of diabetes and changeable antibodies that could be used to treat microbial poisons and snake venoms. In addition to its health advantages for humans and poultry, EYA might be utilised as an alternative to some synthetic antibiotics for animals in order to produce immunological sera, boost productive performance, and increase economic efficiency.

15

Evaluation of Egg Quality Using Non-Destructive Methods

Kiran M, Wilfred Ruban S and Deepika Jamadar

Abstract

To ensure a high and constant egg quality, researchers examined the possibility of replacing the candling process using current sensor technology. Several types of sensors have been created throughout the last decades. Additionally, ultrasonic, magnetic resonance, and electronic nose-based sensors are researched and described.

15.1 Introduction

In addition to serving as a means of reproduction, avian eggs are a staple meal in the diet of humans and include a natural balance of all the required components. However, eggs must overcome a few drawbacks in the current food market in order to compete for sales with an expanding number of other items. For instance, eggs are a delicate commodity that age-related quality reduction. Additionally, the intrinsic variability of the egg's three primary parts—the shell, albumen, and yolk—does not meet the demands of contemporary customers for consistency. The successful sale of eggs depends on planned production techniques and effective quality control processes, both of which contribute to decrease this fluctuation.

15.2 Egg Quality Evaluation

Egg quality may be classified into exterior and internal quality and includes a variety of characteristics linked to the shell, the albumen, and the yolk. Consumer appeal depends on the egg's exterior quality, which is affected by the degree of flaws. Cleanliness, form, texture, and soundness are taken into consideration while judging egg shells. Air cell size, albumen quality, yolk quality, and the presence of blood or meat stains are all indicators of internal quality. The quality of the egg's contents is mostly determined by the albumen. Albumen thinning is a symptom of declining quality. A fresh egg's spherical yolk is in the centre and surrounded by thick albumen when it is delicately cracked onto a smooth, level surface. When a stale egg is cracked, the yolk is flattened and frequently moved to one side, and the thick albumen around it has thinned. As a result, a significant portion of the albumen has collapsed and flattened, creating a broad arc of liquid. This idea underlies the measurement of Haugh units and is still widely applied to assessing albumen quality. Yolk appearance, texture, hardness, and scent are all indicators of quality. Measuring egg quality shouldn't be restricted to processing facilities or to informing consumers. Non-destructive (and destructive) egg quality measurements can be used to grade eggs for improved consumer safety and quality assurance, but they can also be helpful to producers because they provide details on the process's results, which in turn provide details on the flock's health or the climatic conditions in the poultry house. Egg internal properties may be measured non-destructively in order to keep track of the incubation phase.

15.3. Mechanical Techniques

The physical condition of the shell, including its strength and the existence of cracks, has been assessed using mechanical procedures over a considerable amount of time. In general, eggshell strength approaches may be categorised as either direct or indirect. The compression fracture force measured during quasi–static compression is the most prevalently employed metric of a material's strength. Other procedures include puncture and impact testing. These direct approaches are all detrimental. Indirect approaches, both destructive and nondestructive, measure an eggshell strength-related metric. The connection between the various approaches is modest, and the preferred method is frequently determined by the application. Measuring the eggshell's thickness is a common indirect way for determining the eggshell's strength. Calculating the weight % of an egg's eggshell provides an additional indirect indication of the egg's strength. The semi-static compression of eggs between parallel plates is a third indirect and non-destructive technique that is commonly utilised. By measuring the force–deformation curve, the egg's static stiffness may be determined. The slope of the force–deformation line measures the stiffness of the shell. The deformation, often referred to as non-destructive deformation, is the amount by which an eggshell bends or deflects in response to an applied force. It may be used to determine the force necessary to shatter the eggshell and measures a structural attribute of the egg. On the idea that there is a correlation between indirect and direct data, indirect techniques are employed to quantify eggshell strength.

15.3.1 Vibration Analysis

15.3.1.1 Detection of Crack

It has been demonstrated that such rapid, nondestructive mechanical sensors may be utilised to detect cracks in eggshells under commercial conditions. In two commercial applications, an eggshell is agitated by a tiny impactor. As an indicator of the local mechanical eggshell integrity, a combination of the amplitude of the rebounds and/or the number of rebounds of the impactor are utilised. A surface of eggshell that is locally unbroken enables the impactor to conduct several elastic rebound motions with high amplitude. In the vicinity of a fracture, the nearby shell area's elasticity is severely reduced, and as a result, the rebound will be considerably damped. By repeating this process at various locations over the eggshell's surface, a spatial map of the eggshell's mechanical condition may be generated. Either a ball bearing contained within an electromagnetic probe or the egg itself can serve as the impactor.

15.3.1.2 Eggshell Strength

For the detection of eggshell strength, Coucke (1998) modelled the egg as a mass–spring system and proposed an unique eggshell strength metric called dynamic stiffness. The word dynamic was selected to emphasise that the shell strength was determined using dynamic measures, i.e. after being struck, with a correlation between static and dynamic stiffness of 0.71. There found a 0.60 association between eggshell thickness and dynamic stiffness. Later, the basic mass–spring model was expanded to a mass–spring–damper model and it was demonstrated that the damping of the vibration gives additional information. This was determined by correlating the dynamic readings with the static measurements, however its interpretation is still debatable. In addition, they demonstrated that the egg's form is required to connect dynamic and static measures.

15.3.2 Spectroscopic Techniques

15.3.2.1 VIS-NIR Spectroscopy

Visible (VIS) and Near Infrared (NIR) Spectroscopy is a frequently employed method for determining the internal quality of agricultural goods. The examination of NIR absorption spectra yields quantitative data regarding substances such as water and proteins. NIR measurements offer several benefits, including their speed, nondestructiveness, precision, dependability, contactlessness, and affordability. Because touch may be limited for sanitary reasons and a large number of eggs can be evaluated in a short amount of time, the contactless nature of the approach makes scoring egg quality characteristics particularly desirable. Frequently, no sample preparation is necessary. NIR spectroscopy measures the vibrations induced by the stretching and bending of hydrogen bonds with carbon, oxygen, and nitrogen. Experience with these optical methods for determining the quality of eggs is fairly limited.

15.3.2.2 Albumen Quality

Haugh units are still the standard reference technique, however it could be interesting to relate optical or other non-destructive physical measures to a physical albumen quality characteristic, such as viscosity or acidity, rather than to the integrative Haugh unit. There was a correlation of 0.82 between the measured and projected Haugh units, which is satisfactory for classifying eggs. In an effort to reduce the expense of the technique, they demonstrated that the Haugh units could also be predicted using only three optical transmission ratios ranging from the visible to near infrared regions of the spectrum. This may be accomplished with one measurement. Thus, a correlation coefficient of 0.7 was found between the measured and projected Haugh units.

15.3.2.3 Pigmentation of Shell

The colour of the eggshell has little effect on the nutritional content of the egg for humans, but it has a significant impact on customer preferences, making it an essential economic quality criterion. Eggshell colour is mostly influenced by the presence of the pigment protoporphyrin, however the pigment biliverdin also adds to the observed colour of the eggshell. Each colour has a distinct absorption peak. In order to measure the colour of the egg's shell, only light that is reflected from the egg is employed, as pigments within the egg might interfere with light transmission. The colour of the eggshell is then measured by the reflected light at three distinct wavelength bands, which correspond to the wavelengths of red, green, and blue light. Depending on the kind of application, these three parameters are combined to generate a particular colour value, which offers information on the colour of the eggshell. Automated colour grading approaches are not installed on large-scale grading equipment, although they are utilised in layer line selection centres.

15.3.2.4 Blood and Meat Spots

Transmission spectra of eggs allow for the detection of interior optically active abnormalities by shining a light through the egg and analysing the light's spectrum as it passes through. Blood and flesh stains are examples of such irregularities. So yet, there is no meat spot detection method that is available for purchase. Contrarily, machines that grade eggs rely on the ability to identify blood inside the product. Haemoglobin, the blood's optically active pigment has three primary absorption peaks at 415, 541, and 577 nm. Only the 577 nm absorption peak may be utilised to determine the presence of blood in eggs, as all transmitted light below 550 nm is absorbed by the calciferous shell.

15.3.2.5 NMR Spectroscopy

Nuclear magnetic resonance, or NMR, is what happens when the nuclei of some atoms are placed in a static magnetic field and then exposed to a second oscillating magnetic field. This characteristic, called spin, occurs in certain nuclei but not others. A potential non-destructive approach for evaluating the quality of whole eggs is low-resolution nuclear magnetic resonance spectroscopy. Egg freshness might be determined by measuring the longitudinal and transversal relaxation periods, which provide insight into the chemical-physical changes in the item under examination. Researchers found that transverse relaxation periods changed during storage. It was determined that the increased liquification of albumen during storage was to blame for these shifts.

15.3.3 Camera Inspection Systems

For the evaluation of many aspects of egg quality, numerous authors have adopted camera inspection systems. The spatial resolution of the camera, the lighting technique, the number of measurements for each egg, the feature extraction algorithm, the colour of the shell, and the size and kind of defect all have a role in the overall accuracy of camera inspection systems. The processing time required by these camera inspection systems has been cited as a major problem; however, with the ever-increasing capability of current computers, this is seen as less of a barrier, suggesting a rise in the systems' relevance. The existence of transparent sections in eggshells and low-cost alternatives like vibration analysis make them particularly promising for the identification of dirt on eggshells but less so for the detection of cracks.

15.4 Conclusion

Multiple concepts have been developed in response to the need for rapid, non-destructive, and trustworthy methods of evaluating egg quality. Although several of these ideas, like the acoustic or optical technique, have shown promise in laboratory studies, no practical implementations have yet emerged in the commercial sector. One explanation is that conventional techniques have been in use for well over 50 years, thus it is natural for people to trust the values they assign to things like Haugh units and shell breaking force when measuring similar quantities. However, these experiments are damaging and cannot be used on a big basis. People are still hesitant to use phrases like dynamic stiffness of an egg or optical absorption at a specific wavelength, for example. It may not be optimal, however, to present results in the same old standard units that were used for destructive earlier procedures like breaking strength. Combining cutting-edge information technology with a quick and non-destructive quality evaluation tool offers several benefits that extend far beyond the traditional grading stage of the egg production process.

16

Sanitation and Decontamination of Eggs

Deepika Jamadar and Kiran M

Abstract

Sanitary facilities and clean equipment are major elements for the production of safe, high quality eggs and egg products. When spoilage and pathogenic microorganisms remain on product contact and noncontact surfaces at the beginning of the processing shift, there is a greater likelihood of product contamination. Different sanitizing chemicals such as quaternary ammonium, sodium hydroxide, disinfectants, phenols, antibiotics, flumisol, hydrogen peroxide, timsen, poly hexa methylene biguanide hydrochloride (PHMB), formalin, and formaldehyde. Various kinds of decontamination methods are followed in the production of disease-free and hygienic eggs. Chemical decontamination methods involve washing with sanitizers, Electrolyzed water, and Ozone. Whereas, physical decontamination methods are Pasteurization, Irradiation, Microwaves, Ultraviolet light, and Pulsed light.

16.1 Introduction

Successful egg production is highly dependent on sanitation and decontamination, and producing clean eggs is the most major element of egg sanitation. Egg contents might become contaminated either vertically or horizontally. The ideal option for small farms can be to keep hatching eggs in a hygienic area and incubate them as soon as possible without sanitation but it's not the same kind in case for commercial producers need to get a poultry veterinarian guidance on the proper egg handling and sanitation program depending on their needs. There are several preharvest techniques used to minimize egg contamination, including immunization, feed additives, home sanitation, bacteriophages, and biosecurity. Because the contamination of shell eggs cannot be totally eliminated by these preharvest techniques, postharvest strategies have been explored as a means to reduce or eliminate the contamination of shell eggs. The aim of this chapter is to discuss on major sanitation and decontamination methods of shell egg.

16.2 Egg Sanitation

There are several ways to sanitize hatching eggs. If a farm needs to be sanitized, the cost of alternative equipment and chemicals as well as the size of the enterprise and potential uses for the chicks will all be important considerations. Traditional methods of sanitization include washing with water or immersing in disinfectant solutions. There have been numerous studies accomplished to look into various chemicals, such as egg sanitizers, including quaternary ammonium disinfectants, antibiotics, sodium hydroxide 1-2%, hydrogen peroxide (H_2O_2) 1.4-1.5%,flumisol, phenols, timsen, formalin as a 0.5% solution PHMB 0.035%, formaldehyde applied by fumigation.

There are steps followed for production of cleaned and sanitized eggs.

- Dry cleanup –Removes soil and dirt
- Rinse – Using warm water that is less than 120 °F, rinse remnants off equipment and floors. An increase in the water's temperature can make proteins stickier and more difficult to remove.
- Detergent application and scrubbing – Application of approved cleaning chemicals for complete removal of protein and fat. Foaming agents or Manual scrubbing of surfaces help with scrubbing action is recommended.
- Final rinse – Get rid of any detergent residue (Hot water used).
- Inspect and spot-clean – Ensure that hard-to-reach places on and within equipment and surfaces are checked. Hard-to-reach areas should be the target for this step.
- Sanitize –sanitizing chemicals used to reduce bacterial load. (Fumigation and Spray Application)
- Air-dry.

16.2.1 Disinfectants for Egg Sanitation

Washing table eggs using chlorine-based disinfectants containing a cleaning agent which has proven to be safe for hatching egg sanitation. For washing hatching eggs, a number of commercial disinfectant formulations are registered. According to studies from the University of California, quaternary ammonium is a superior sanitizer for hatching eggs. Quaternary ammonium has the benefit of being safe. The advantages of quaternary ammonium are that it is reasonable cost, safe for hatching eggs, and equipment, leaves residual protection on eggs

16.3 Regulations of Egg Sanitation

Two distinct food safety organisations monitor the two main segments of the egg industry. The U.S. Food and Drug Administration (FDA) regulates shell eggs, whereas the USDA Food Safety and Inspection Service governs egg products (USDA-FSIS). To meet food safety standards, each of these organizations has its own set of guidelines. While USDA-FSIS regulations are based on the Hazard Analysis and Critical Control Points (HACCP) system, FDA regulations are based on preventive control techniques. Both of these approaches are designed to decrease the threat to the public's health caused by diseases like Salmonella.

16.4 Methods of Decontamination of Shell Eggs

16.4.1 Washing

Hatching eggs can be sanitized by egg washing if the right equipment are available to perform accordingly. The wash water's suggested temperature range is 110° to 120°F (43.3° to 48.9°C), which must be hotter than the eggs. However, washing could contaminate the eggs if the water temperature is below the recommended level or if contamination exceeds the disinfectant's capacity (a particular concern in immersion washers or reservoir-type). The washing solution must contain a sufficient amount of sanitizer. No more than 200 eggs should be washed in a gallon of fluid when using an immersion washer, and the entire washing time should not exceed three minutes. Reservoir-type washers need to have mechanisms to monitor and control sanitizer levels in order to function at their best. There are a number of commercial egg washing devices that, when used properly, successfully sterilize hatching eggs. For washing eggs, only high-quality water with less than 2 ppm iron should be used.

16.4.2 Electrolyzed Water

A weak salt water solution is electrolyzed to create electrolyzed oxidizing water (EOW), which has an acidic pH of 2.6 and an alkaline pH of 11.4. EOW has been shown to be effective at killing germs on a solid surface, in food, and in liquids. Since alkaline EOW has a high pH, it can be used as a substitute for detergents with a high pH, but acidic EOW has a low pH and a high oxidizing reduction potential, making soluble chlorine an efficient sanitizer.

16.4.3 Pasteurization

The technique of pasteurization involves applying heat to liquid foods to eradicate microorganisms. Pasteurization of eggs using heat has received less attention from researchers than the commonly used pasteurization of liquid whole eggs, egg whites, and yolks. Salmonella may be eliminated from several food products using heat.

16.4.3.1 Hot Air Pasteurization

Eggs were immersed in a circulating water bath set at 57°C for 25min, which resulted in a 3log reduction of the inoculated Salmonella. A hot air oven set at 55°C for 180 minutes, however, resulted in a 5log reduction in *S. Enteritidis*. Pasteurization techniques, including the use of a water bath or a hot air oven, and inoculated intact shell eggs with *S. Enteritidis*. Combining the two techniques heating the water bath at 57°C for 25 min and then heating the hot air for 60 min at 55°C produced a reduction in Salmonella of 7 logs. Following all treatments, the quality of egg whites was found to be satisfactory.

16.4.3.2 Microwave Pasteurization

Use of microwaves, which heat the food from the inside out. Theoretically, the albumen should heat faster in a microwave than yolk, and this was confirmed in studies with the individual components. However, trials with intact shell eggs demonstrated that in actuality the yolk and albumen heated at a similar rate.

16.4.4 Thermoultrasonication

Utilization of ultrasound to clean bacterial spores of debris the spores were subsequently more sensitive to heat treatment than were spores that did not receive the ultrasound treatment. Later, many authors discovered that using both heat and ultrasound to combat vegetative cells was more successful than using either one alone.

16.4.5 Gas Plasma

A gas that has been ionized and contains equal numbers of positively and negatively charged particles is called gas plasma. By exposing a gas to an electric field, create a gas plasma that contains free radicals and charged particles that can interact with vital components of the bacteria to disturb their metabolism.The procedure is practical and affordable to implement because of the usage of atmospheric pressure and low temperatures, and it has certain applications in healthcare.

16.4.6 Pulsed Light

Pulsed light processing uses broad-spectrum high-intensity light in short bursts to kill microorganisms and has applications in food packaging, processing equipment, and medical devices. Dunn (1996) inoculated shell eggs with *S. Enteritidis* and subsequently treated the inoculated eggs with pulsed light; he determined that the treatment was able to eradicate the inoculum from the shells of the eggs.

16.4.7 Ozone

Ozone (O_3) is a very strong antimicrobial agent active against microorganisms at relatively low concentrations. Ozone has been researched for potential uses in the food industry, and it has been approved by the US FDA for use as an antimicrobial in foods (FDA, 2001). Ozone is unstable and decomposes quickly to oxygen and is also effective at low temperatures.

16.4.8 Irradiation

For fresh shell eggs, the FDA authorized the use of ionizing radiation up to 3 kGy in 2000. (FDA, 2015). Additionally, numerous studies evaluated the sensitivity of five *S. enteritidis* isolates inoculated on or inside of whole shell eggs to gamma radiation. The inoculated eggs were exposed to radiation at dosages of 0, 0.5, 1.0, or 1.5 kGy. All of the isolates were eradicated from the surface of the eggs with a dose of 0.5 kGy, however, the sensitivity of the isolates inside the eggs differed significantly.

16.4.9 UV Light

- UV light exposure can be used to sterilize eggs. New non-thermal, non-destructive methods like UV-C light treatments and hybrid techniques like UV light and ozone and UV light and hydrogen peroxide were developed.
- UV light treatment is now a trustworthy, non-thermal alternative to conventional treatments for egg shells and a good choice for preventing microbial contamination of the surface of the egg shells. The following are the most relevant findings from the analysis of the research done to date are:
- In comparison to methods that used chemical sanitizers, UV light treatment significantly reduced the bacterial population of clean and recently contaminated egg shells;
- UV light treatment facilitates the hygiene control of the transportation system
- Elimination of microorganisms could be achieved at low temperatures and under relative dry conditions by using UV light treatments;
- UV light treatment ensured significant reductions of Salmonella spp., the most frequently present pathogen reported in eggs;

- *S. typhimurium* treated with UV light on egg shells did not recover after subsequent incubation under either dark or light conditions;
- *Y. enterocolitica* was more UV light resistant than the native aerobic microbiota of the egg shells;
- UV light treatment of hatching eggs in a prototype cabinet can effectively reduce the aerobic and pathogenic bacteria on egg shells.
- Salmonella was effectively inactivated on egg shells in a short period of time and at low temperatures with the use of a combination of UV light and ozone treatment, or UV light and H_2O_2 treatment; however, UV light treatment does not reduce the internal contamination of eggs due to the impermeability of eggshells to UV radiation.

17

Layout and Design of Egg Processing Plant

Rajendra Kumar K

Abstract

Plant design refers to the overall design of a manufacturing enterprise / facility. It moves through several stages before it is completed. The stages involved are : identification and selection of the product to be manufactured, feasibility analysis and appraisal, design, economic evaluation, design, report preparation, procurement of materials including plant and machinery construction, installation and commissioning. The design should consider the technical and economic factors, various unit operations involved, existing and potential market conditions etc.

17.1 Introduction

The plant layout needs a proper arrangement of each constituent area, along with their interrelationship. One of the most important considerations is the future expansion potential. The plant layout design should permit expansion readily and economically. One side of the building should be free of permanent installations such as compressors, septic tanks, toilets, drain fields, offices, and loading docks to permit the addition of building units without relocating these auxiliary facilities.

Other factors to consider in the overall plan include

1. Easy access to related plant areas,
2. The ease with which the layout can be rearranged,
3. Access to the cooler for incoming eggs,
4. The layout of roadways and parking areas,
5. Ease and simplicity of materials handling operations,
6. Working conditions and employee satisfaction,
7. The ease with which management can supervise all operations, overall appearance, and utilization
8. Adaptability to local conditions (topography, climate, access roads, and community conditions).

The layout for the areas should be arranged in a straight line, with the processing room in the center and storage areas on each end. The cooler should be located close to the point where eggs are graded and packed eggs accumulate, and the dry storage should be adjacent to the point of packing materials use. Offices, facilities requiring plumbing, and fixed equipment are placed at the front of the building, and the loading dock is placed at the cooler end. This type of arrangement leaves one part of the plant free for future expansion and provides easy access to the processing area from all areas within the plant. Since most of the activities and employees are in this area, the convenience of the area is a prime consideration. To reach the employee lounge, employees must use the office entrance passage. Thus employee movement can be monitored and wall space will not need to be broken up.

17.2 Site Selection

The location of the plant was a serious problem for some plants (when located adjacent to production operations) because areas around the site had become heavily populated and complaints were registered with local governments about offensive odours. Such problems can be forestalled by selecting the site carefully. Several plants had inadequate parking space, while others had poor accessibility for trucks and poor drainage.

The site selected must provide sufficient space for an efficient arrangement of the overall facility. Under ideal circumstances, space should be adequate to drive completely around the structure within the property boundary lines. The site should have adequate room for development, employee parking, and for off-street management of trucks.

1. The selection of the site is influenced by
2. The initial cost of the land,
3. Clearing and grading the site,
4. Construction of roads,
5. Binging in the utilities,
6. Fencing or landscaping the property,
7. Cost of transporting eggs and packing materials to and from the plant,
8. Availability of labor, and
9. Prevailing rates for taxes and insurance.

The relative proximity of the egg supply should also be considered, since locating the processing operation at or near the production site is desirable. In such a location, the plant must be located in proper relation to the prevailing wind and pedestrian traffic routes to prevent possible contamination of the flock. When plants are located in rural areas that have no sewage treatment facilities, the cost of disposing of waste from the egg washer in a system separate from the sanitary sewer must be considered.

17.3 Design of Plant and Auxiliary Facilities

The first aspect of the design to be considered is the height of ceilings. The height of the ceiling should not be considered individually. However, all features in the structural design of a building should be studied by their interrelationships with all other features that may be involved in the functional efficiency of the building. In the design for the ceiling and roof system, the interior should be free of obstructions will be considered. A clear interior contributes to the efficiency of work and ease of maintenance. The ceiling height of 14 feet is recommended for the stacking of pallet loads and for storage facilities that have been developed around a forklift operation.

The efficiency of operation, ease of maintenance, and durability are the functional requirements in building use. A good start on the functional requirement is made by building the shelter only. The building, or shelter, is simply put around and over an operation that is laid for the efficiency of the operation.

The ease of maintenance is the second functional requirement; it contributes to general efficiency, safety, and pleasant working conditions. Cleaning of the storage areas should be considered in designing the plant. Sloped floors make stacking difficult, especially high stacking with forklifts, and are not recommended in storage; areas. Walls and partitions between storage and other areas should be of smooth construction to reduce dust problems. One or more trapped drains located in storage areas permit ease in cleaning- by sweeping water into the drains.

Frequent cleaning of the processing area is required more than other areas because of the nature of the operation and the intense use of limited floor space. Cleaning floors presents a little problem since drains for machine wastewater are usually available and conveniently located. It is recommended that the exterior walls be placed on curbs that extend 6 inches above the floor. This arrangement facilitates cleaning and reduces water contact with the wall lining. An impervious and easily cleaned wall surface may be applied over the interior lining. It should meet the changing building use or more stringent sanitation requirements. Electrical and heating equipment, as well as walls and other structural features, should be made of the material that will permit the occasional use of hoses to clean the processing area.

Durability under the conditions of service in an egg packing plant, the third functional requirement, is largely protection against physical damage by materials handling and equipment. The masonry wall in the cooler door is framed with steel angles to prevent damage from the forklift. To prevent damage from contact with the pallets or the forklift trucks, the base of the wall is mounted with a 4 by 4-inch wood member. The damage to the doors is prevented by covering them with a heavy metal sheet.

The handling of materials being carried to or dragged away from the building is an important aspect of functional design. A dock at the truck-bed level is most desirable. The dock may be raised (pit) type, or it may be a raised platform. An aboveground-level platform is a good approach where a cut-and-fill situation exists. The environment usually dictates the type of loading dock selected.

The selected loading docks must absorb the shock when trucks strike the dock, drainage must be provided for the pit of raised docks and care must be made for the difference in the angle of the truck body as it affects the contact point of a truck with the dock. Proper care must also be made for the maneuvering of trucks into the dock area.

17.4 Roof and Wall System

Stud walls with a structural plywood lining and corrugated metal siding complete the building. The interior lining was used as a structural member in the design of the building, as being the most economical approach, since a complete lining was considered essential in the processing and cool storage areas. In addition, it permits ease in cleaning and maintenance. When the interior lining is used as a structural member, a variety of materials from which to select an exterior covering is possible. Since the exterior covering should be as maintenance-free as possible, the greater freedom of choice of materials can provide additional economy in construction.

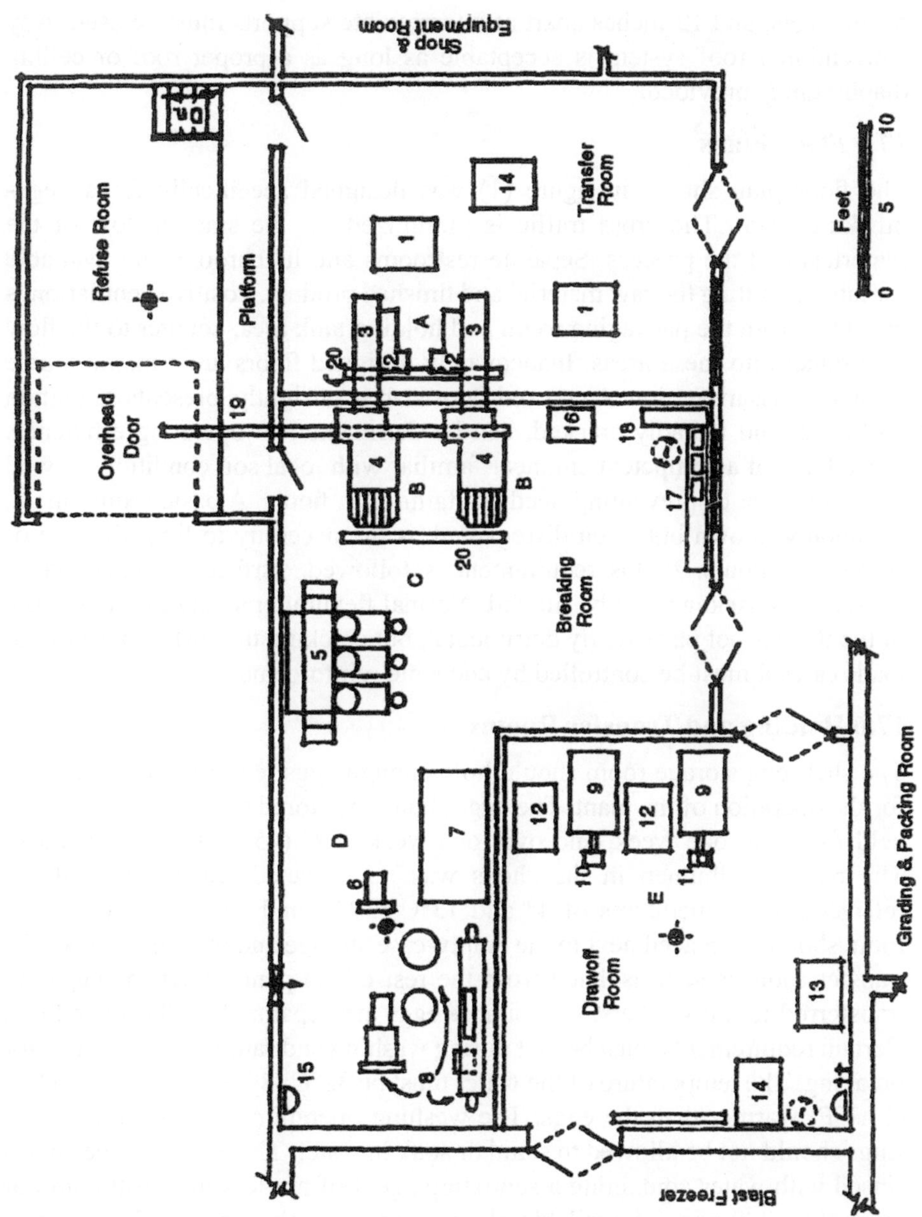
Shop &
Equipment Room
Refuse Room
Platform
Overhead
Door
Transfer
Room
Breaking
Room
Drawoff
Room
Grading & Packing Room
Blast Freezer
Feet
0
5
10
A
B
C
D
E
1
2
3
4
5
6
7
8
9
10
11
12
13
14
15
16
17
18
19
20

The plywood nailing schedule should be observed as closely as possible in the development of diaphragm action. For both the roof and wall lining, 8mm nails spaced 4 inches apart at wall and roof boundaries, 6 inches apart at other panel edges, and 12 inches apart at intermediate supports must be used. Any conventional roof system is acceptable as long as a proper roof or ceiling diaphragm is provided.

17.5 Floor Plans

The floor plan shown in figure (1) was designed specifically for an egg-breaking plant. The cross traffic is minimized for the smooth flow of the materials and the process. Separate restrooms and lunchrooms are available for labor handling the raw material and finished product. Positive ventilation is provided from the packaging room and holding tank area, counter to the flow of product into these areas. Inadequately designed floors are a major source of owner dissatisfaction. A cracked floor slab is unsightly, presents sanitation problems, and if badly cracked, can be detrimental to operating efficiency. The advice of a competent engineer familiar with local soil conditions is well worth the cost of preventing needless failures in floors. A basic requirement, commonly known but often disregarded, is the necessity to limit the size of continuous pours. If this requirement is followed, surface cracking that is caused by shrinkage will be limited. Normal flexural cracking on the bottom of the floor is not particularly detrimental, but cracking the surface because of loads carried must be controlled by adequate reinforcement.

17.6 Holding and Transfer Rooms

The shell-egg storage room should hold enough eggs required for a few days for the operation of the plant. The eggs should be stored at 13°C (55°F), when held for up to one week, and over one week, 7°C (45°F) is recommended. The residual albumen in the shells was not affected when stored at the refrigeration temperatures of 4° and 13°C (39.2° and 55.4°F). The storage room should be placed next to the empty-case storage and transfer rooms. The transfer room is semi-isolated from the rest of the plant. After the eggs are transferred to the washers, undesirable eggs are separated by flash candling. Certain requirements must be met during washing and sanitizing shell eggs for breaking. The temperature of the water must be 32°C (89.6°F) or at least 11°C (11.7°F) warmer than the eggs. The washing operation must be continuous. Eggs should not be allowed to stand or soak in water. The eggs must be spray-rinsed with water containing a sanitizing agent of not less than 100 ppm nor more than 200 ppm of available chlorine or its equivalent. Shell eggs must be sufficiently dry at the time of breaking to prevent contamination from free moisture present on the outside of the shell.

17.7 Cold Storage Requirements

The exact space requirements for cooler storage are difficult to determine. Space must be availablse for incoming eggs (ungraded); outgoing finished product (graded and packed eggs); in-transit items such as wire baskets, filler flats, and empty pallets; and any eggs held for breaking if in-plant egg-breaking facilities are to be included in the space requirements..

17.8 Breaking Room

The breaking room should have adequate lighting, ventilation, unscented soap, warm water, hand washing stations operated by means other than hand controls facilities, clean towels or other facilities for drying hands and the person involved in the breaking operation shall have personal hygiene, which is essential for the breaking room operation. A minimum of 30-foot candles of light are required in all working areas in the breaking room, except at inspection stations and where the shell eggs are broken, which require 50-foot candles of light. The packaging of liquid egg products is done in separate rooms with filtered positive air ventilation. The breaking equipment should be cleaned and sanitized before use and shall be cleaned at the end of each shift. Tanks and churns shall be cleaned every four hours. Breakers must use a complete set of clean equipment when starting work and after lunch periods.

Whenever an inedible egg is broken, the contaminated equipment must be replaced. This practice shall reduce bacterial counts and yolk contamination of egg white. Packaging films and containers used for packaging should not pass through the breaking room or areas where shell eggs and cases are handled. Tanks, vats, drums, or other containers used for holding liquid eggs must be of approved construction, fitted with covers, and located in rooms maintained in sanitary condition. The temperature requirements for liquid-egg products after breaking are shown in Table 17.1.

Table 17.1: Temperature requirements for liquid egg products

Product	**Unpasteurised product temperature within 2 hr from time of breaking**			**Temperature within 2 hr after pasteurization**	**Temperature within 3hr after stabilization**
	Liquid (other than salted product) to be held 8 hr or less	**Liquid (other than salted product) to be held in excess of 8 hr**	**Liquid salt product**		
Whites (not to be stabilized)	12.8°C(55°F) or lower	7.2°C(45°F) or lower	-	7.2°C(45°F) or lower	-
Whites (to be stabilized)	21.1°C(70°F) or lower	12.8°C(55°F) or lower	-	12.8°C(55°F) or lower	-
All other product (expect product with 100% or more salt added)	7.2°C (45°F) or lower	4.4°C (40°F) or lower	-	If to be held 8 hr or less, 7.2°C (45°F) or lower if to be held in excess of 8 hr, 4.4°C (40°F) or lower	If to be held 8 hr or less, 7.2°C (45°F) or lower if to be held in excess of 8 hr, 4.4°C (40°F) or lower
Liquid egg product with 10% or more salt added	-	-	If to be held 30 hr or less, 18.3°C (66°F) or lower if to be held in excess of 30 hr, 7.2°C (45°F) or lower	-	-

17.9 Breaking Machines

The development of egg-breaking and separating machines was a major improvement in the efficient production of liquid-egg products. The first units had a capacity of about 10,000 eggs/hr but the machines are currently highly sophisticated mechanical, automated, and computerized machines operating at up to 65,000 eggs/hr. The following functions are performed: automatic transferring (loading); visual inspection and washing (including sanitizing) of shell eggs; cracking the shells, separation of white, yolk, and shells, inspection of contents; and removal of white, yolk, and shells. The equipment parts (usually stainless steel) coming into contact with the product can be removed, cleaned, and sanitized as needed. These variable speed machines can handle a wide range of egg sizes. One to five operators with scanners run them. Edible eggs with broken yolk membranes are diverted into liquid whole eggs. Loss or rejection becomes an inedible liquid.

17.10 Packaging Room

The edible liquid products are filtered; ingredients are added, standardized, blended, and pasteurized as needed. A separate room is needed for packaging to avoid recontamination of the pasteurized product. Many types of packages are used. The packaging materials must not pass through the breaking room where unpasteurized products exist. Cross traffic should be avoided. All the personnel in breaking and packaging rooms must wear caps or hair nets properly.

Detailed floor plan of breaking plant

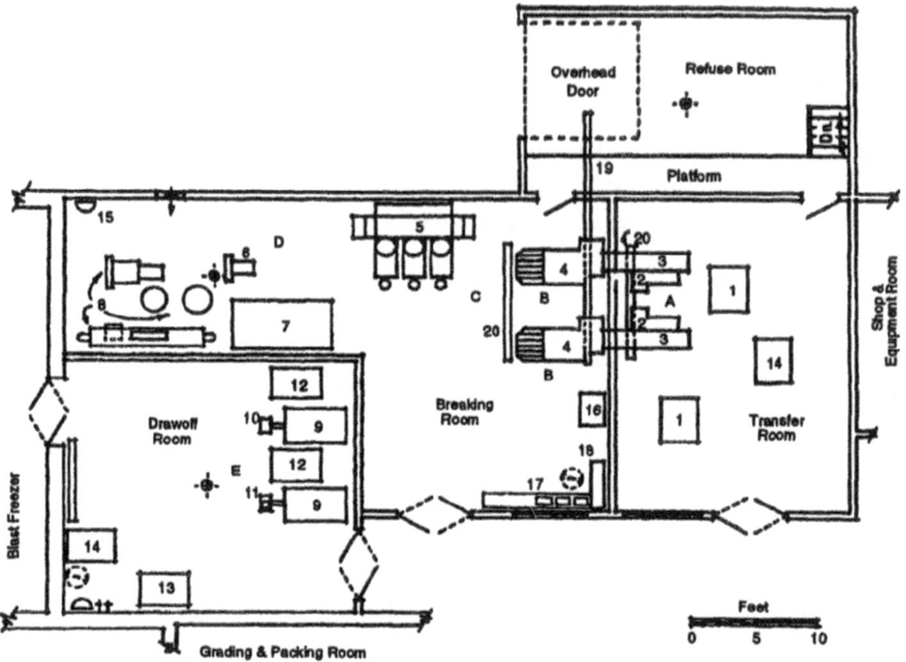

Work Stations	
A. Loader	8. Pasteurizer & Accessories
B. Machine Operator	9. Bulk Cooler Tank
C. Attendant	10. Plastic Bag Filler Unit Scale
D. Foreman	11. Pallet of Egg Meat
E. Draw off room	12. Pallet of Empty Containers
Equipment	13. Empty Pallets
1. Pallet of Eggs	14. Lavatory (foot-operated)
2. Case Bench	15. Table
3. Egg Conveyor	16. Sink (3 compartments) with drain rack
4. Egg-Breaking Machine	17. Rack for Soiled Utensils
5. Chums & Strainers	18. Auger
6. Plate Cooler	19. Gutter Drain
7. Bulk Cooler Tank (3 compartments)	

Large quantities of liquid egg products are now packaged in aseptic containers (which have an extended refrigerated shelf life) for institutional uses. These packaging systems require strict sanitation and adequate pasteurization practices which start with top quality shell eggs. Frozen products shall be reduced to -12°C (10.4°F) or lower within 60 hr from the time of pasteurization.

17.11 Shell Room

The empty shells are conveyed to this room and usually centrifuged to remove adhering white. Also, eggs removed by candling (inspection) are centrifuged. This inedible liquid can be pasteurized and then sold for pet food. The shells are discarded as garbage, spread on fields as fertilizer or dried, and then recycled into animal feed. Other uses are in plastics, concrete, cosmetics, bathroom cleaners, or toothpaste.

17.12 Requirements for an Egg Grading and Packing Plant Layout

In planning and preparing an efficient layout for the egg grading and packing plant, the following points should be considered.

1. The travel distance to cooler and dry storage areas should be minimum.
   - Minimize cross-flow of product and packing materials in high-activity areas.
2. Provide ample space for positioning the pallets with a forklift truck.
3. Provide sufficient space around machines for cleaning, servicing, and repair.
4. Provide space for all auxiliary equipment.
5. Allow space for machine modification.
6. Provide an arrangement that is compatible with the equipment and facilities needed for expansion.
7. Locate the floor drain to allow for flexibility in placing equipment.
8. If rooms are added, they should be designed so as not to extend into the processing area.

Install utilities, where they will permit future modification of facilities- without major structural change.

17.13 Office and Employee Facilities

The office, restrooms, and breaking room and freezer, if a breaking operation is considered should be located on one side of the building. This arrangement will facilitate the future expansion of the building and the common use of plumbing and sewage facilities. In addition, it will minimize space-wasting projections into the processing area and possible interference with flow patterns. The building arrangement should provide an entry into the front of the building and limit, to one point entry into the processing area. The time clock should be placed at the entrance to monitor the entry of employees into the rest area and the entrance of visitors. The same person can also handle over-the-counter sales and observe the operation of the processing area.

The general office provides space for two desks, a counter, and office equipment. The adjoining private office and restroom are optional. If the sales, record maintenance, and accounting are to be done at the plant, additional space (optional private office) should be given for the general office work.

17.14 Personnel Facilities

The employees should be provided with comfortable surroundings in their workstations and in the rest areas. The design of these facilitates better worker morale, greater productivity, and improved workmanship. The restrooms adjoining the employee lounge should be separated by a hall that can be utilized for employee lockers, water heaters, storage space for supplies used in the area, and cleaning materials. The restrooms are provided with an exhaust fan or with windows opening to the outside.

17.15 Wastewater

The BOD (biological oxygen demand) of wastewater from egg-breaking plants ranges from 1,000 to 22000 mg per litre (Harris and Moats, 1975). Egg solids are introduced into the wastewater from broken eggs during washing and from the cleaning of equipment. These researchers were able to reduce the BOD by 76-97% in a process that involved acidification to pH 4.7 with H2SO4, heating to 75°C (167°F), and then removing the precipitated proteins by filtration or centrifugation. The disadvantages of this process (Sievers, Vandepopuliere, and Harris, 1983) are high equipment costs and the fragility of the floc. Removal by foam formation was more promising. Obviously, the best solution to a high BOD of wastewater is to reduce its source during processing. Another problem in wastewater is caused by cleaners and sanitizers. A slug-flow of an alkaline cleaner can cause high pH levels after clean-up. The use of sanitizers can also cause sewage treatment problems.

18

Poultry Egg Waste

Deepika Jamadar and Kiran M

Abstract

Eggs have been considered one of the high-nutritional-value food for humans, and they are widely consumed worldwide. Several hundred egg product factories process approximately 1 million eggs daily. This production generates extensive volumes of eggshells and egg trays, which has resulted in a significant waste-handling challenge. There are various kinds of uses of poultry egg waste is not only in the food industry but also seen in non-food industries such as the medicinal industry, as an adsorbent, as fertilizer, as a supplement and in Biodiesel production, as a coating pigment and Bio-filler.

18.1 Introduction

The egg industry has grown quickly in the last decade as people's eating habits have become more aware of the importance of good and nutritious diets. Eggs are now widely accepted as a good source of high-quality protein. However, at the same time, the eggshell that pollutes the environment is discharged into the open. This eggshell waste can be used to make a variety of items, including pharmaceuticals and medicines. Additionally, this raw material is inexpensive and conveniently accessible. Eggshells are ranked as the fifteenth biggest environmental pollutant. Recycling these industrial eggshell wastes is essential for the sustainable growth of our earth. Small drops can become mighty oceans, so even if this didn't significantly contribute to the end of pollution, it was still a positive step. Egg shell fragments can be quickly and effectively turned into useful materials for industrial uses rather than being discarded as waste.

18.2 Applications of Egg Waste

18.2.1 Medicinal Use

Egg shells : Bioactive components found in egg waste are advantageous for medical applications. In terms of this aspect, the manufacture of biocompatible materials or organic matter from them has introduced a new dimension to the usage of Egg Shell agricultural waste. Used eggshells remain useless and unconverted. Calcium phosphate was found to be important as a biomaterial due to its oleophilic nature and its inclusion in bone tissues. calcium carbonate which is a significant component of bones and bone utilized in dentistry is present in eggshells in levels as much as 94%. Additionally, reusing eggshells to produce (HAp) lowers waste pollution and seeks to implement cleanliness while also converting that waste into a highly valuable product that minimizes the risk of bone repair or therapy modification. The synthesis of HAP and Nano-HAp from eggshell waste is an ecofriendly technique.

Eggshell Membrane

The Natural Egg Shell Membrane (NEM) is a novel dietary supplement that comprises naturally occurring glycosaminoglycan and proteins that are required to cure pain and elasticity caused by joint and connective tissue abnormalities.A number of studies have assessed natural eggshell membranes as a source of glucosamine, hyaluronic acid, chondroitin, and collagen and revealed that it is an effective, safe therapeutic option for treating pain and elasticity problems associated with joint and connective tissue disorders. This has led to the rejuvenation of medicines made using it for patients suffering from Arthritis or other joint pains.

18.2.2 Adsorbent

Many researchers have tried to assess the usage of chicken eggshell waste as a natural and affordable absorbent to remove hydrogen sulphide from wastewater. It destroys aquatic life because by loweringthe amount of dissolved oxygen in the water. The eggshell is taking the lead in hydrogen sulphide absorption. The physical and chemical characteristics of processed eggshells were investigated for the purpose of using it as an absorbent, and the potential utility of removing their heavy metals from synthetic and actual wastewater. It has been proven that natural eggshell waste can effectively extract chromium, cadmium, and lead from wastewater. Silver metal has been successfully extracted from technical trash using an eggshell-based adsorbent solution.

18.2.3 Fertilizer

Every day a large number of eggshells are produced around the world as biowaste. It can be used as a fertiliser for plants since it guards against the disease known as blossom-end rot (BER) of the plant and helps in lowering farming costs. In general, the eggshell waste that is left over from industrial processing is utilized in agriculture just from being crushed and ground into a fine powder to repair soils with an acidic pH. Its application as fertiliser improves plants' ability to absorb minerals and salts from the soil. Additionally, because eggshells are a rare natural source of calcium carbonate, using them as an alternative option can reduce the impact on limestone resources.

18.2.4 Biodiesel Production

Waste eggshells can be utilized to produce calcium oxide that is used as a heterogeneous catalyst in biodiesel products. Calcium carbonate dominates the chemical composition of eggshells. One of the most essential oxides for biodiesel synthesis is calcium oxide, which can be recycled more than 8 times without losing any of its usefulness. The eggshell is a useful raw material for producing thin powder because of its abundance and the numerous interior pores on its surface. In fact, the catalytic calculation procedure results in the creation of solid bases derived from eggs.

The idea of recycling eggshell waste was introduced with the aim of producing biodiesel in a sustainable manner, as reducing the amount of contaminants, lowering the cost of production, and reusing eggshell trash as a low-cost catalyst. The scientists achieved 100% biodiesel production by using 4% weighted eggshell waste as a catalyst for the trans-esterification of cooking oils for 5 hours. In addition, this calcium oxide catalyst can be reused more than 8 times without a reduction in its working capacity.

18.2.5 Feed Supplement

The process of extracting calcium from oyster shells is an expensive process because oyster shells are not readily available. Therefore, chicken eggshell appears to be a suitable alternative source of calcium supplement, as it is easier to assemble and easier to extract from the eggshell than calcium from oyster shells, and it is also highly soluble.Eggshells areentirely made up of calcium carbonate, which makes up your nails, teeth, and bones, as well as Proteins, Magnesium, Selenium, Strontium, and other compounds that keep bones and joints healthy. The nutrients found in the eggshell are easily digested and get absorbed in your body and the best part is that they are 100% free.

There are many calcium supplements made from eggshells available in the market and you can also make calcium supplements from eggshells at home after proper channel of processing. Poultry breeders can use eggshells exclusively as a source of Calcium in the diet of male chickens, as it has no detrimental effects on weight gain, semen quality, reproductive development, and calcium homeostasis.

18.2.6 Coating Pigment

Corrosion has been a problem for so long in carbon steel pipes (the most common and versatile engineering material used in pipes and ducts for water transportation) and coatings are applied to protect it from corrosion. Calcium carbonate obtained from the eggshell can be used as a coating pigment in Carbon steel to protect the pipes from rust. Also using the eggshell as a coating pigment for an inkjet printer provides another way of its reuse. Previously, synthetic pigments based on silicon oxide were considered suitable in inkjet printers but they had some advantages and disadvantages. The silicone used to have high image resolution, high color intensity, and fast-drying but it was not cheap and it was in high demand for construction. Therefore, eggshells proved to be an effective alternative to Silicon oxide as they are readily available and inexpensive. The use of eggshells as a coating pigment for inkjet printers improved the optical density of dye-based inks and reduced the gloss density of black ink and paper. Therefore, eggshells can be an effective source for their use in inkjet printers with multi-color prints instead of simple black and white prints.

18.2.7 Bio-filler

Linear low-density polyethylene (LLDPE) is an important polymer product used to process useful products such as containers, sheets, table coverings, etc. It is used in various applications due to its chemical neutrality, toughness, anti-corrosion, and anti-cracking flexibility. Due to its excellent qualities, it is extremely desired but expensive. To solve this problem and improve its physical and chemical properties, cheap inorganic materials are used as fillers in low-density polyethylene. Calcium oxide and hydroxyapatite derived from eggshells have been found to be beneficial in improving its physical and chemical properties, as well as being widely beneficial as a bulk filler as it is inexpensive, lightweight, and little weight bearing. Moreover, the adhesive properties in the eggshell are higher than in other components used as biofilters.

18.2.8 Handicrafts

There are many experiments using eggs or eggshells that are great learning experiences for kids at home. In this, the naked egg experiment is where you will completely dissolve the eggshell and then you can separate thin layers one by one of the eggs when you will finally see the egg yolk and white. Eggshells are the perfect material to make mosaic art because it breaks down into really interesting shapes. You can paint your shell with different colors and then it can be used for all kinds of crafts -for photo frames, pictures on paper, and much more. Creating Easter or Christmas decorations using eggshells is fun for kids or adults too. And it's also pretty easy to make; you do not need to boil the eggs. You will only make two small holes in your egg- one at each end - and remove the egg yolk and white. Then you can decorate them with your fancy Easter theme, Christmas theme, or in fact anyway.

19

Egg Quality

Kiran M

Abstract

Egg quality refers to various standards that define both external and internal quality. The internal quality is focused on the yolk height, yolk color, Albumin Viscosity, and Haugh unit. In contrast, the external quality refers to the eggshell thickness, egg width, height and cleanliness. Egg quality is a general term which refers to several standards which define both internal and external quality. External quality is focused on shell cleanliness, texture and shape, whereas internal quality refers to egg white (albumen) cleanliness and viscosity, size of the air cell, yolk shape and yolk strength.

19.1 Introduction

Quality has been defined as the properties of any given food that have an influence on the acceptance or rejection of this food by the consumer. Egg quality is a general term which refers to several aspects of both internal and external quality. Internal egg quality involves functional, aesthetic and microbiological properties of the egg yolk and albumen. The proportions of components for fresh egg are 32% yolk, 58% albumen and 10% shell. The quality of an egg is at its best when the egg is laid. There is no known method by which this quality can be improved subsequently. Hence eggs have to be maintained in fresh condition as far as possible; otherwise their quality will deteriorate rapidly.

19.2 Internal Egg Quality

The egg white is formed by four structures. Firstly, the chalaziferous layer or chalazae, immediately surrounding the yolk, accounting for 3% of the white. Next is the inner thin layer, which surrounds the chalazae and accounts for 17% of the white. Third is the firm or thick layer, which provides an envelope or jacket that holds the inner thin white and the yolk. It adheres to the shell membrane at each end of the egg and accounts for 57% of the albumen. Finally, the outer thin layer lies just inside the shell membranes, except where the thick white is attached to the shell, and accounts for 23% of the egg white. Egg yolk from a newly laid egg is round and firm. As the egg gets older, the yolk absorbs water from the egg white, increasing its size. This produces an enlargement and weakness of the vitelline membrane; the yolk looks fl at and shows spots. As soon as the egg is laid, its internal quality starts to decrease: the longer the storage time, the more the internal quality deteriorates. However, the chemical composition of the egg (yolk and white) does not change much.

19.3 External Egg Quality

To maintain consistently good shell quality throughout the life of the hen, it is necessary to implement a total quality management programme throughout the egg production cycle. Exterior egg quality is judged on the basis of texture, colour, shape, soundness and cleanliness according to USDA (2000) standards. The shell of each egg should be smooth, clean and free of cracks. The eggs should be uniform in colour, size and shape.

19.4 Factor Affecting Egg Quality

Quality has been defined as the properties of any given food that have an influence on the acceptance or rejection of this food by the consumer. Egg quality is a general term which refers to several aspects of both internal and external quality. Internal egg quality involves functional, aesthetic and microbiological properties of the egg yolk and albumen. The proportions of components for fresh egg are 32% yolk, 58% albumen and 10% shell. The quality of an egg is at its best when the egg is laid. There is no known method by which this quality can be improved subsequently. Hence eggs have to be maintained in fresh condition as far as possible; otherwise their quality will deteriorate rapidly.

19.4.1 Damage to the Egg Shell

The egg shell immediately gets cracked if the egg is subjected to blows, jerks, jolts etc. Cracks in the egg expose the inner contents to unfavorable environmental conditions, objectionable odors and invasion of micro-organisms. Moisture and carbon dioxide can also escape through the cracks. Weak shells, rough handling, poor packing conditions, rough journey etc., increases the chances of damage to the shell.

19.4.2 Evaporation of Moisture

Moisture starts evaporating from the inner contents of the egg and vapours escape through the shell pores soon after the egg is laid. It continues later on also. The rate of loss of water due to evaporation, however, depends upon the temperature, humidity and the movement of air around the egg and the porosity of the egg shell.

High environmental temperatures, low humidity and rapid air movements around the egg and a higher porosity of the egg shell quicken the loss of water, while low temperatures, a high humidity and still air around the egg and poor porosity of the shell slow down the same.

The loss of water from the egg shell results in

1. Decrease in the weight of the egg.
2. Increase in the size of the air-cell. This is due to the shrinkage of the contents and subsequent drawing of air into the egg.
3. Decrease in specific gravity of the egg.

The loss in weight without corresponding loss in the volume of the egg results in a decrease in its specific gravity. The extent of loss in the weight, the decrease in the specific gravity of egg and the increase in the size of the air-cell will depend upon the amount of water lost through evaporation.

19.4.3 Chemical Disintegration

The albumen and the yolk undergo decomposition if the egg is not stored under proper conditions. High environmental temperatures are particularly conducive to these changes. In an ageing egg, the proteins break down, fats decompose and certain other chemical reactions involving the calcium carbonate of the shell take place. As a result of these changes, ammonia, volatile oils, water, hydrogen sulphide etc., are produced in the egg. Undesirable odours and smells develop in the egg and its quality progressively goes down till it becomes inedible.

19.4.4 Liquefaction of Thick/dense Albumen

High environmental temperatures cause the disintegration of the mucin fibres of the thick albumen which then gets liquefied. This change means a major deterioration in the egg quality. The liquefaction of the thick albumen is also affected by the concentration of carbon dioxide in the atmosphere surrounding eggs in the egg store.

19.4.5 Escape of Carbon Dioxide

In a newly laid egg, the albumen of the egg contains carbon dioxide. Subsequently, this gas is continuously released from the egg due to chemical reactions taking place in the albumen and the yolk and on the calcium carbonate of the shell. Higher environmental temperatures quicken these reactions and increase the production of carbon dioxide and its escape from the egg. This adversely affects the quality of the egg contents. This process is also affected by the concentration of carbon dioxide in the atmosphere surrounding the eggs.

19.4.6 Flattening and Floating of the yolk

The vitelline membrane which separates the yolk and the albumen is semi-permeable. In a fresh egg, the yolk is more concentrated than the albumen. Due to difference in osmotic pressures on the two sides of the vitelline membrane, water flows from the albumen through this membrane to the yolk. Consequently the viscosity of the yolk decreases and its volume increases. The vitelline membrane stretches due to the increase in volume of the yolk material, and gets weakened as the pressure from inside goes on increasing. The yolk tends to flatten as its water content increases. The yolk floats in the albumen due to the liquefaction of the dense portion. When this happens, the yolk tends to move away from its central position towards the shell.

Stuck yolk – When eggs are left in a fixed position for a long time and the dense albumen gets liquefied, the yolk floats towards the shell and gets attached to the membranes. This condition is called Stuck yolk.

Seeping yolk- If a stuck yolk is loosened from its position, the vitelline membrane usually breaks and permits the yolk material to seep into the albumen. The first stage of this condition is usually called Seeping yolk.

Addled egg – When the yolk material mixes with the white, as a result of rupturing of the vitelline membrane, the condition is usually called Addled egg or Mixed rot.

19.4.7 Changes in Reaction

With the ageing of the egg, the alkalinity of the albumen increases and the acidity of the yolk decreases. The changes are mainly due to the escape of carbon dioxide from the egg. High environmental temperatures quicken these changes. Escape of carbon dioxide from the egg, liquefaction of the thick albumen, flow of water from albumen to yolk and changes in their reaction can be slowed down by storing eggs at low temperatures and by adding carbon dioxide to the atmosphere of the egg store.

19.4.8 Phospholipids Decomposition

There is a progressive increase in the ratio of inorganic to total phosphorus when the eggs are stored. Phospholipids decompose to liberate glycerophosphoric acid as well as inorganic phosphoric acid. The vitelline release inorganic phosphoric acid, and the phospholipids into white, thereby the phosphorus content of the white increases.

19.4.9 Absorption of Objectionable Odours

In addition to the production of odours inside the egg due to chemical disintegration and decomposition, the egg contents also readily absorb odours from the environments. Eggs stored with onions, garlic, kerosene oil or any decomposing matter pick up their unpleasant odours and go down in quality. Eggs may pick up a musty odour from the nesting material (in the laying nests) or packing material in egg containers. This odour may also be produced by the action of certain bacteria. This defect can be found out by smelling either the packing material or the egg-case immediately after opening the latter, or by breaking open the eggs and examining the odour of their contents.

19.4.10 Growth of the Embryo

The optimum growth of the embryo takes place at about 99.5-99.75° F. At this temperature, blood is produced at the end of two days exposure. As the development of the embryo proceeds, the quality of the egg contents deteriorates. The egg becomes inedible as soon as blood appears in the embryo. If the death of the embryo takes place before blood has been formed, a red spot appears at the place of germinal disc, over the surface of the yolk. This spot is called Heat spot. An egg with a heat spot is edible, though it is of a lower quality. If the death of the embryo occurs shortly after blood has been formed, a small red ring is produced at the surface of the yolk. This is called Blood ring. If the embryo dies at an advanced stage of development, decomposition sets in immediately and the egg rots. Eggs having blood ring or dead embryos are inedible.

19.4.11 Bacterial Decomposition

The contents of a newly laid clean egg are usually free from contaminating organisms. The safeguard from a subsequent entry of bacteria is provided by the partial sealing of the pores by the layer of the cuticle over the shell. Before laying, the oviduct's content of antimicrobial agents and secretion of the egg white proteins seems to provide an effective barrier against the microorganisms present in the cloaca. Antiperistalsis may carry forward the microorganisms from the cloaca. But when the eggs are stored in unsanitary environments, their shells get laden with organisms. Salmonella is the known bacteria voided through the faeces as well as through the eggs from infected/carrier birds. The bactericidal property of albumen declines with the ageing of the eggs, especially if they are stored under high temperatures. When these changes take place, the contents of the egg form a good medium for the bacterial growth, which is further stimulated by high environmental temperatures. The bacterial decomposition of the egg is accompanied by the disintegration of yolk and albumen and the formation of by-products of decomposition. In other words, the egg begins to rot. The egg contents become discoloured and emit a rotten odour. The yolk liquefies and albumen becomes watery and turbid. Various types of Rots (Rotten eggs) produced by the bacterial action in the egg are as given in Table 19.1.

Table 19.1 : Various types of common egg rots produced by bacterias

Types of Rot	Organism
Black rot type 2	*Proteus spp*
Black rot type 1	*Aeromonas liquefaciens*
Custard rot	*Certain enterobacterers*
Red rot	*Serratia marcescens*
Green rot	*Pseudomonas maltophilia*
Pink rot	*Pseudomonas fluorescens*
Fluorescent green rot	*Pseudomonas putida*
Fluorescent blue rot	*Pseudomonas aeruginosa*
Yellow rot	*Flavobacterium cytiophaga*
Colourless rot	*Other Enterobacters, Alcaligenes*

19.4.12 Moulds/Fungi

A high humidity in the environments favour the growth of fungus on the egg shell. The fungus is soft and white and popularly known as Whiskers, grows on the shell when the relative humidity in the cold store is 95 per cent or more. This fungus does not penetrate into the egg. But many other fungi penetrate through the shell and grow either on the inner surface of the shell or in the albumen and yolk. On the inner surface of the shell membrane, the fungi produce small yellow, blue, black or green spots. In the albumen, they produce coagulated masses around them. Further growth results in complete gelling of albumen and breakdown of the yolk membrane. Wrapping the eggs in a wet cloth or keeping them on a wet surface for a long time moistens the shell and facilitates the entry of fungi through its pores. The fungal growth inside the egg can be detected by candling.

19.4.13 Mottling of the Egg shell

Dry air quickens the loss of water vapour from the inner contents of the egg and also the accompanying changes in the albumen and yolk. When eggs are stored in a dry atmosphere, more moisture is drawn from the inner contents through the pores of the shell. A large portion of this moisture escapes into the atmosphere, and a small amount gets spread unevenly in the shell which acquires a mottled appearance. The mottling of the shell, by itself, does not mean deterioration in the quality, but it may indicate improper storage conditions and the accompanying lowered quality.

19.5 Prevention of Egg quality Deterioration

To minimize egg quality problems two things are important: frequent egg collection, mainly in the hot months, and rapid storage in the cool room. The best results are obtained at a temperature of 10 °C. There are six main factors affecting internal egg quality: disease, egg age, temperature, humidity, handling, and storage.

1. ***Disease:*** Newcastle disease, EDS-76 and infectious bronchitis produce watery albumen, and this condition may persist for long periods after the disease outbreak has been controlled.
2. ***Egg age:*** eggs several days old show weak and watery albumen, and the CO_2 loss makes the content alkaline, affecting the egg flavour.
3. ***Temperature:*** high temperatures cause a rapid decrease in internal quality. Storage above 15.5 °C increases humidity losses.
4. ***Humidity:*** high relative humidity (RH) helps to decrease egg water losses. Storage at an RH above 70% helps to reduce egg weight losses and keeps the albumen fresh for longer periods of time.
5. ***Egg handling:*** rough handling of the eggs not only increases the risk of breaking the eggs, but also may cause internal egg quality problems.
6. ***Storage:*** eggs are very prone to take on the odours of other products stored with them; separate storage is therefore advised.

20

Egg Quality Deterioration

Kiran M, Deepika Jamadar and Wilfred Ruban S

Abstract

The following Tables in chapter summarize factors that may affect egg quality and suggest corrective measures. As more emphasis is placed on egg quality, it is important that all possible defects be eliminated. When defects are found, consult the chart for possible causes and solutions.

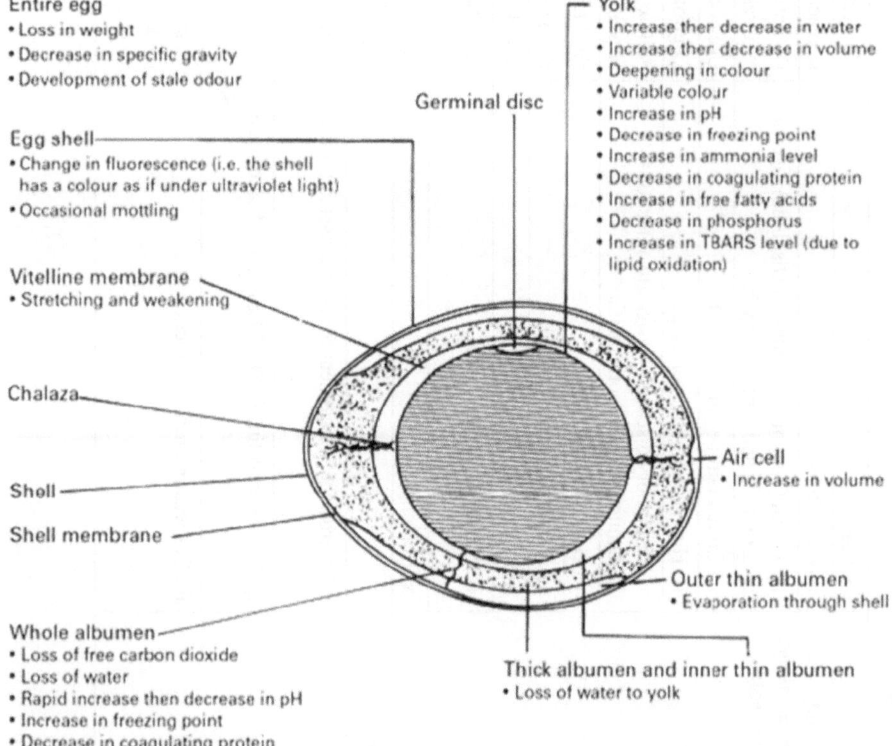

Figure 20.1: Quality deterioration in egg due to aging

Table 20.1: Egg Shell

Condition	Causes	Corrective Measures
1. Thin, sandy, misshapen, rough, ridged or soft	i. Age of hens	Replace after 12 to 14 months of lay.
	ii. Arasan (Tetrame Thylthiuram Disulfide); used to control mold and fungus growth.	Do not include arasan treated grains in the diet of layers. Law requires that treated grain be dyed.
	iii. Sulfanilamide (sulfa drugs)	Use according to accepted recommendations
	iv. High constant temperatures	Control temperature. Provide plenty of water.
	v. Respiratory diseases (Newcastle, infectious bronchitis and laryngotracheitis)	Follow a recommended program for vaccination and disease prevention in poultry.
	vi. High salt (NaCI)	Feed less salt.
	vii. Drugs for rodent control	Keep rat bait away from poultry.
	viii. Fright	Avoid sudden noises. Approach birds cautiously
	ix. Reduced calcium intake	Provide 3 percent calcium during cool weather. Provide 4 percent calcium during warm weather.
	x. Heredity	Select strains that produce eggs of good shell
2. Abnormal color		
a) Brown to yellow	i. Nicarbazine; a treatment for coccidiosis in broilers.	Do not feed to layers.
	ii. Chlortetracycline (600-800 gm./ton), Aureomycin	Use according to accepted recommendations
	iii. Gas lead in lines or burners	Maintain tight connections in gas line, keep burners in excellent working condition and provide ventilation in areas where gas is used to heat egg rooms.

b) White to brown	Iron ($FeSO_4$), ($7H_2O$) 0.1 ppm	Have iron content checked in water used for washing eggs. Keep below 0.1 ppm.
c) Faded color	Low calcium in diets of turkeys.	Raise calcium levels for increased shell color and improved shell thickness.
d) Mottling of shell (bright spots or moist appearance around pores, observed by candling)	Water retained by protein in spongy layer of the shell.	Do not mistake for a crack or cracks in the shell. Maintain 80% humidity in egg room.
3. Porosity	Age and breed of hens, environmental temperatures, and season of year.	Keep hen house cooler, hold eggs in cool place, sell hens after 12 to 14 months of lay or molt, and select strain bred for good shell texture.
4. Tremulous or moving air cell (observed by candling)	Rough handling	Observe and make necessary alterations in egg- handling practices.
5. Tainted shells	Paradichlorbenzene (moth repellant)	Do not feed to birds.

Table 20.2. Egg White

Condition	Causes	Corrective Measures
1. Pink egg whites	Cottonseed meal (often found in cattle rations)	Avoid using in the diet of layers
2. Weak, thin or watery whites	i. Age of hens	Replace hens after 12 to 14 months of lay.
	ii. Ammonia from droppings	Better ventilation, use superphosphate on litter and manure and remove droppings regularly.
	iii. Increased alkalinity, (pH)-Loss of CO_2	Use a shell coating such as oil or refrigerated temperatures (40 to 55°F).
	iv. Respiratory diseases (Newcastle, infectious bronchitis, and laryngotracheitis)	Follow a recommended program for vaccination and disease prevention in poultry.
	v. Heredity	Select strains of known egg white (albumen) quality.
	vi. Arasan	Do not use arasan treated grains in the diet of layers.
	vii. Vanadium	Use sources of phosphorus in feeds known to have low amounts or none.
	viii. High environmental temperatures	Collect eggs often (three to five times a day) and hold in refrigerated temperatures (40 to 55°F.).
	ix. Sulfanilamide (sulfa drugs)	Use according to accepted recommendations.
3. Flecks or spots in albumen	i. Partially cooked	Avoid excessive heat when washing eggs
	ii. Blood and meat spots	Select strains known for clear egg whites (albumen).

4. Green rot and other types of microbial spoilage.	Microorganisms, including bacteria, molds, and fungi	Maintain clean nesting materials. Gather eggs frequently (three to five times a day). Use clean water for washing eggs. Maintain temperature of egg wash water (100 to 120°F.) above that of the egg at all times. Use recommended amounts of detergents and sanitizers. Keep equipment clean. Use clean packing materials. Keep eggs refrigerated. Green rot is easily detected with an ultraviolet lamp candler. Other types of advanced spoilage are easily detected with regular candling techniques. Egg wash water containing 0.4 ppm of iron can promote bacterial spoilage.
5. Cloudy white	i. Prompt oiling of newly laid shell eggs.	Delay oiling for one to six hours after eggs are laid.
	ii. Prompt refrigeration of newly laid shell eggs at 32 to 39°F.	Keep eggs refrigerated below 45°F.
6.Off-odors and flavors	Chemicals for treating parasites. Odorous flowers, fruits, and vegetables in egg storage areas.	Use chemicals recommended for lice and mite control. Do not use materials capable of imparting odors or flavors to eggs such as BHC, Lindane or Hexaphene. Do not store flowers, fruits and vegetables in the same area with eggs.
7. Blood and meat spots	i. Hemorrhaging before and during ovulation.	Tranquilizers, vitamins A and K, and aureomycin
	ii. Breed	Select strains with low incidence.
	iii. Continuous intermittent periods of light.	Use 14 hours of light.
	iv. Color or pigment caused by porphyrin as found in the brown shell egg.	Select strains with low incidence

Table 20.3. Egg Yolk

Condition	Causes	Corrective Measures
1. Olive- or salmon-colored yolks	5% or more cottonseed meal (found in some cattle diets).	Avoid its use in the diet of layers. Do not allow freerange poultry to consume spilled cattle feed.
2. Platinum yolks (colorless yolks)	Possible infection (causative agent unknown)	Antibiotics. (200 gm auremycin and 2 lbs NF-180 per ton of feed for seven days.)
3. Colorless yolks	Lack of xanthophyll	Consideration should be given to the source of xanthophyll such as yellow corn meal, alfalfa leaf meal, etc.
4. Green yolks	i. 100 to 250 mg of sodium chlorophyllin in feed.	Avoid feeding to hens.
	ii. Seed pods of Shepherd's purse and pennycress	Use clean grains in feeding programs
5. Greenish-brown yolks	5 gm or more of pimiento peppers daily to each hen	Use smaller amounts for a desirable color in egg yolks.
6. Orange-pink yolks	Red pepper	Avoid feeding to hens.
7. Yellow to orange yolks	Seaweed meal (algae), dehydrated alfalfa meal, corn gluten meal, flower petal meal, dried chili peppers, powdered African red peppers, dried sweet potatoes, dried carrots, corn oil products, food grade fat soluble dyes, etc.	Feed recommended levels of xanthophyll bearing materials for desired egg yolk color. Yellow =13 mg of xanthophyll per lb of feed Medium orange = 23 mg of xanthophyll per lb of feed Orange = 34 mg of xanthophyll per lb of feed. Maximum color will be present 10 days after the hens are placed on feeds for yolk color.
8. Misplaced egg yolk	Large end up with yolk in large end – thin egg white and/or fat content of yolk. Large end up with yolk in small end – thin egg white and/or water content of yolk.	Use accepted quality control practices while gathering and storing eggs in a cooled atmosphere.
9. Blood and meat spots	Hemorrhages (ovarian, may be inherited)	Select strains with low incidence. Older breeds have approximately 30 % blood spots in eggs.

10. Mottled or blemished yolks	Nicarbazine	Do not feed to layers.
	Cottonseed meal	Avoid feeding to layers.
	Piperazine citrate	Do not use frequently or continuously.
	Movement of water from egg white across vitelline membrane into yolk material	Cool eggs quickly and keep cool. Use other accepted quality control practices.
11. Thick, pasty, rubbery or cheese-like	Crude cottonseed oil (malvalic acid and sterculic acid)	Avoid feeding to layers.
	Yolks laid internally	Remove offending birds from the flock
12. Apparent misplaced egg yolk (observed in the whole egg by candling)	Unknown	The egg positioned with the small end down may help correct this situation.
13. Off-odors and flavors	Chemicals for treating parasites. Odorous fruits and vegetables in egg storage area.	Use of chemical recommended for the lice and mite control. Do not use materials capable of imparting odors or flavors to eggs such as BHC, Lindane or Hexaphene. Do not store flowers, fruits, vegetables or petroleum products with eggs.
	Chemicals or egg washing compounds	Do not place egg-washing powders or liquids directly on eggs.
14. Flat	Weak vitelline membrane	Gather eggs often (three to five times a day). Maintain temperatures of 40 to 55°F. Market often.
15. Stuck yolks	Newcastle disease	Use recommended vaccine. Storage at high temperature Store at 40 to 55°F.

21

Egg Quality Evaluation

Kiran M, Wilfred Ruban S and Deepika Jamadar

Abstract

Eggs have been considered one of the high-nutritional-value food for humans, and they are widely consumed worldwide. Several hundred egg product factories process approximately 1 million eggs daily. Egg quality refers to various standards that define both external and internal quality. The internal quality is focused on the yolk height, yolk color, albumin viscosity, and Haugh unit. In contrast, the external quality refers to the eggshell thickness, egg width, and height and cleanliness.

21.1 Introduction

The egg industry has grown quickly in the last decade as people's eating habits have become more aware of the importance of good and nutritious diets. Eggs are now widely accepted as a good source of high-quality protein. Increased demand in consumer choice has resulted in a wide variety of egg selection available in the retail market. Eggs of high quality are needed to sustain niche markets for specialty and designer eggs–seeking consumers. Egg quality defines those characteristics of an egg that affect consumer acceptability and preference. The parameters for egg grading include shell cleanliness, strength, texture, and shape; the relative viscosity of the albumen; and the shape and firmness of the yolk. While the consumer is concerned about broken or damaged eggshell during purchase, more importantly is the interior egg quality, which begins to wane after egg laying. Efficient egg gathering, cooling, and refrigeration at appropriate humidity can maintain this feature. To evaluate quality, several parameters are used to determine the consistency in egg size and content, as well as contribute to grading of the egg. Furthermore, the structural integrity of an egg's shell and the yolk membrane integrity are directly related to food safety and possible economic loss.

21.2 External Quality of Eggs

21.2.1 Egg Size/Weight

A normal chicken egg weighs 55-60 g depending on breed and age. The chicken egg will be about 1/30th of the hen's body weight.In all species older birds lay heavier eggs than younger birds.

21.2.2 Egg Shape

The usual egg shape is ovate. The shape of the egg plays a major role in packing and transport. Egg shape is express as Shape index (SI). A normal egg will have shape index of 72 (Range 70-74). If the SI is above or below the normal range cannot be hold in egg trays. Egg which is spherical in shape will have a shape index of 75 and above. Those eggs which are elongated /elliptical will have lesser shape index of 70. Measure the diameter of the egg at the broadest point along the short axis of egg. Measure the length of the egg along the long axis. The measurements can be made accurately using Vernier Calipers to an accuracy of 0.01 cm. Calculate the shape index using the formula.

Shape Index (S.I) = (Diameter of the short axis/Length of the long axis) x 100

21.2.3 Shell Colour and Texture

It indicates smoothness and roughness of shell surface and also indicates shell quality. Shell colour is due to the presence of pigments. Ooporphyrin gives the brownish colouration to the egg shell which is normally seen eggs laid by the Asian, the English and the American class of birds.

21.2.4 Cleanliness

This is essential for consumer satisfaction and also to improve and maintain the keeping quality.

21.2.5 Volume

Volume is one of the indicators of egg size. Egg volume is directly proportional to egg size. The egg volume of different species will be around 90-95% of their fresh egg weight. To measure the volume of the egg, fill a measuring cylinder of 500 ml / 1 litre capacity with a known quantity of water. After noting the lower meniscus of the water gently slant the measuring cylinder and slide the egg carefully into the measuring cylinder and note the final reading of the water, the difference in volume will give volume of egg in cm^3.

21.2.6 Specific Gravity

Volume is one of the indicators of egg size. Egg volume is directly proportional to egg size. This gives an indication of egg shell quality, as well as its freshness. Fresh eggs will have higher specific gravity (range 1.0 to 1.1 with average of 1.06) than old and long stored eggs. Eggs having stronger shell will have a specific gravity of 1.06. Any value less than this may indicate that the egg is old or the eggs are thin shelled. Estimated by measuring the egg weight and then weighing the egg in water, to find the weight loss in water. Measured by dipping the eggs in a salt solution having several concentrations of salt dissolved in it, having a specific gravity ranging from 1.0 -1.1 with an interval of 0.02.

Specific Gravity = Weight of the egg in gram/Volume in cc

21.2.7 Surface Area

Surface area of an egg is directly proportional to egg size. The surface area will be more for elongated eggs than for spherical eggs.

Surface Area = 12.6 x{(length +width)/4}

21.3 Internal Quality of Eggs

The quality of egg can be best be ascertained only by breaking open the egg and studying the various parameters of the shell, the albumen the yolk.

21.3.1 Shell Thickness

1. Shell thickness usually ranges between 0.35 to 0.45 mm. commonly about 4.0mm. Below 0.33 mm is generally considered as poor shell quality.

Procedure

1. Take three pieces of shell from broad, middle and narrow ends of egg
2. Remove shell membranes attached to it
3. Measure the Least Count (L.C) of given Screw guage. It is generally 1/100 mm that is 0.01mm.
4. Fix the shell piece between the two jaws of the screw gauage
5. Measure the Pitch Scale Reading (PSR) and Main Scale Readings (MSR).

   Shell thickness = PSR +MSR x L.C

21.3.2 Albumen Index

1. The firmness of the egg white is correlated with the albumen quality. A fresh egg will have an albumen index of 0.1.

Procedure

1. Check the level of the egg breaking stand with the help of spirit level and maintain a perfect level by adjusting the legs.
2. Break open the egg and pour the contents on the egg breaking stand.
3. Immediately record the height of the thick albumen at the highest point using the Spherometer with the least count of 0.01 mm (0.02 mm in some).
4. Using a Vernier callipers record the mean width of thick albumen by taking observations along the long and short axis.
5. Calculate the albumen index using the formula

   Albumen Index: Height of the thick albumen(mm)/Mean width of the thick albumen(mm)

21.3.3 Haugh Unit (H.U)

It is the most widely used value to measure the albumen quality. It is modified version of albumen index, with the height of the thick albumen adjusted to the egg weight by using the Haugh unit meter or by using slide role.

Procedure

1. Weight the egg accurately to the nearest gram
2. Break open an egg over the flat surface of the egg breaking stand.
3. Record the height of the albumen following the same procedure and precautions as described in the determination of albumen index.
4. Calculate the Haugh unit score of the egg using the formula

   $H.U = 100 \log (H+7.57 -1.7 W^{0.37})$

21.3.4 Yolk Index

This is an expression of the spherical nature of the yolk. The average value for a fresh egg is 0.40 and above. As the egg becomes aged, it flattened and the yolk index is lowered.

Procedure

1. Check the level of the egg breaking stand with the help of spirit level and maintain a perfect level by adjusting the legs.
2. Break open the egg and pour the contents on the egg breaking stand.
3. Record the height of the yolk using the Sperometer with the least count of 0.01mm (or 0.02 mm in some).

4. Using a Vernier callipers record the mean diameter of yolk.
5. Calculate the yolk index using the formula.

   Yolk Index=Height of the yolk in mm/Average diameter of yolk in mm

21.4 Candling of shell eggs

1. For proper candling, the egg should be held at about elbow level to allow for good vision. The egg is place at the candling aperture with the air cell up wards and tilted slightly away from the observer.
2. The egg is then quickly twirled or rotated to allow the contents to move. This motion enables to distinguish the yolk, chalaza, blood spots, meat spots etc.
3. Usually two eggs are held in each hand , one between the thumb, forefinger and the middle finger and the other in palm , held by the other two fingers.
4. The first one is candled and slowly glided back to the palm through a rotary motion, while the other one brought up, in position for candling.
5. In this way, the right and left hands are used alternately enabling four eggs to be candled in quick succession.
6. Using both hands, with two eggs in each is a rather difficult job, however greater proficiency and accuracy is obtained with practice and experience. While doing simple candling work often one hand with two eggs or both hands with one egg each are used for candling.

21.4.1 External Quality Factors

21.4.1.1 Shell Shape and Texture

1. *Practically Normal:* A shell that is approximately the usual shape and that is sound and free from thin spots. Ridge and rough areas that do not materially affect the shape and strength of the shell are permitted.
2. *Abnormal:* A shell that may be somewhat unusual or seriously misshapen, faulty in soundness or strength and showing pronounced ridges and thin spots will be downgraded.

21.4.1.2 Soundness of Shell

1. *Sound:* An egg with unbroken shell
2. *Check or Crack:* An egg that has a broken shell or crack in the shell but with its shell membranes intact and the contents do not leak. Checks may range from a very fine hair like crack that is visible only by candling.

1. *Smashed:* An egg whose shell is smashed or shattered.
2. *Leaker:* An egg that has a crack with a break in the shell and shell membranes to the extent that the egg contents leak out.

21.4.1.3 Shell Cleanliness

1. *Clean:* A shell that is free from foreign material and from strains. An egg may be considered clean if it has only very small piece of strains and dirt.
2. *Moderately stained:* A shell which is free from prominent stain and adhering dirt, but has moderate stains.
3. *Dirty:* A shell which is stained prominently beyond 1/4th of its total surface area. A dirty egg will be downgraded.
4. *Adhered dirt:* shell which have foreign dirty material stuck especially faeces etc. This will be totally downgraded.

21.4.2 Candled Out Egg Quality

21.4.2.1 Air cell depth

A fresh egg will have an air cell depth of less than 3 mm. As the egg ages, the depth of the aircell increases.

2. *Practically regular:* In this, the air cell maintains a fixed position seen at the broad end of the egg. The outline of the air cell is clear in a white shell egg.
3. *Free air cell:* An air cell which moves freely between the intact membranes towards the uppermost point of the egg . This will downgrade the egg.
4. *Bubbly air cell:* A ruptured air cell and inner cell membrane resulting in one or more small separate air bubbles usually floating in the albumen.

21.4.2.2 Yolk

1. *Outline indistint or slightly defined:* The yolk outline is indistinctly indicated and appear to blend into the surrounding white as the egg is twirled in a fresh egg.
2. *Outline fairly well defined:* The yolk outline is visible but not clearly outlined as the egg is twirled.
3. *Outline plainly visible:* The yolk outline is clearly visible as a dark shadow when the egg is twirled.

The terms used to describe the yolk defects are:

1. *Practically free from defects:* The yolk shows no blastoderm development many show other very slight defects on its surface.

2. *Serious defects:* The yolk may show well developed germ spots and other serious defects.
3. *Clearly visible germ development:* Development of the blastoderm in the yolk that has progressed to the point where it is plainly visible , as a circular spot on the yolk with no blood.
4. *Blood due to germ development:* Blood carried by development of the blastoderm in a fertile egg to a point where it is visible as a definite line or blood ring. These eggs are discarded.

21.4.2.3 Albumen

1. *Clear:* The albumen is free from discoloration of foreign bodies floating in it. Prominent chalazae should not be confused with meat spots.
2. *Thickness:* Fresh egg will have firm thick albumen which keep the yolk in central position. As egg becomes old, the albumen will become thin and watery and the yolk will be freely moving.
3. *Meat spots and blood spots:* May be on the surface of the yolk or floating it is greater than 0.3 cm (1/8 inch) diameter if it is greater than 0.3cm ($1/8^{th}$ inch) diameter, the egg is discarded.
4. *Bloody white:* Blood has diffused through the white. The egg cannot be consumed hence discarded.

22

Marketing Channels and Integration

Sushant Handage and Kiran M

Abstract

In recent decades there has been a sustained and substantial shift in human diets across the globe towards including more livestock-derived foods. However, there remain unanswered questions about how demand for livestock-derived foods may evolve over the upcoming decades for a range of scenarios for key drivers of change including human population, income, and consumer preferences. Thus, marketing of livestock products is not simply in looking for the channels but also adequately studying the consumer preferences.

22.1 What is Livestock Product Marketing?

Livestock Products Marketing is the change of ownership of livestock derived food products. It links the livestock production and food consumption. There are three aspects of livestock market transactions:

1. Spatial – transactions occur across space
2. Temporal – transactions occur across time
3. Form – transactions occur in a certain form

Thus, livestock marketing involves various business activities that direct the flow of livestock goods and services from producer to the consumer at the appropriate time, place and in the form the consumer desires.

22.2 Facets of Livestock Marketing

1. Coordinating the process of exchange: How you reach the food to the people who demand it? Here both the suppliers and consumers are part of market and marketing.
2. Geographical aspects: Locations where buyers and sellers meet.
3. Value addition: This includes the transformation, transportation and storage aspects

22.3 How Is Livestock Market Suffering in Comparison to the Traditional Market?

1. Seasonality affects the demand.
2. Risk of perishability – risk that good will become unusable. Meat and eggs are perishable items
3. Weather and other unpredictable events
4. Potential for high price volatility: The prices of inputs and outputs are volatile. Increasing cost of feeding and decrease in market price for the outputs is always a risk.
5. Competitive markets (price taking environment): Livestock market is an unorganised market infrastructure. supply and demand determine prices; producers accept these prices when they sell, and consumers accept these prices when they buy
6. Lack of forward integration.

22.4 Demand for Most Farm Products is Inelastic

People can consume only so much then they are satiated. Even if price drops, they will not buy much more. When demand is inelastic a drop in price that spurs more quantity being sold results in lower revenue and profit for the producer. This has always been a challenge for livestock producer.

22.5 Channels of Livestock Marketing

22.5.1 Direct Channel

Producer – Consumer

Online retail: Many farmers and producers are now selling eggs and meat online through their own websites or through online marketplaces like Amazon and Swiggy.

Social media: Many farmers and producers use social media platforms such as Facebook, WhatsApp business and Instagram to promote their eggs and meat products and connect with potential customers.

22.5.2 Indirect Channel

Producer –Collector–Assembly Merchant – Wholesaler – Retailer – Consumer; Wholesale distribution: Eggs and meat can also be sold to wholesalers, who then distribute them to restaurants, hotels, schools, and other food service providers.

Producer – Food Service providers – Consumer: Eggs and meat can also be sold to food service providers, such as schools, hospitals, and prisons, for use in their meals.

The efficiency of marketing channel is reflected by the share of the consumer's rupee received by the producer. The higher the share, greater the efficiency of marketing channel. Egg prices vary from one market to another and from one season to another. In summer, wholesale egg prices were low to the level which is sometimes lesser than the cost of production. Therefore, proper attention has to be given to efficient disposal of market eggs.

22.5.3 Integrated Market

Integration is the association, coordination, amalgamation of companies engaged in various stages of production of particular product, or related products, so that, there will be a smooth flow of inputs and outputs from one unit to other, leading to overall reduction in the cost of production of the final product.The integrators provide the day - old chicks (doc), feed, medicine, veterinary services, and management guidance and are also responsible for removing and marketing the mature birds. The farmer provides the house and equipment to the integrator's specification, power, fuel, labour, and day-to-day management. Bonuses are most commonly awarded for exceeding contractual performance bench marks for mortality and FCR. In the south the integrator pays about 3 Rs per kg of harvested bird based on an FCR of 2.0 and mortality 4.0 percent, plus 0.50 Rs.per Kg in incentives for lower FCR or mortality. In the south contract growing of broilers is well established.

Due to integration the retail prices are lower in the south; the poultry integrators in addition to reducing the production cost, have sharply reduced marketing margins between producer and retail prices. The process of integration has created regional oversupply conditions that has lowered retail prices and squeezed trade margins. The integrators have replaced the wholesalers and have established their own retail presence and, in the process, have changed the sector from a high margin, low - volume business to a low - margin, and high - volume business. This process could be attributed to the expansion plans that each integrator has planned and this entails the required help from the contractors to be loyal to them. Upward 75% of production in the south is now reported to be integrated, much more than in other regions.

The layer and parent stock farmers will receive 20 weeks old pullets/ parent stock and feed from the integrator and supply table/ hatching eggs to the integrator on commission basis. The integrator will clean, spray egg coating oil, grade, pack and distribute the table eggs to various marketing channels. The hatching eggs are utilized to produce pullet chicks for the next cycle. Since, the culled hens are not used for table in developed countries, they are mainly utilized for pet food manufacturing and the offal are recycled in poultry feed, after converting it into meat meal, feather meal and chicken fat.

22.5.3.1 Vertical Integration

It is the association between different stages of production starting from a base operation such as breeding, hatchery, commercial farm, processing unit and retailing in order to utilize the output of one unit as input of the other unit. This will ultimately reduce the cost of production of the final product, namely the egg and meat, without affecting the profit margin of the integrator.

22.5.3.1.1 Forward Integration: This is a vertical integration that moves forward. Here, in order to fetch a better price for his output / product, the integrator will start his own processing plant or marketing centre.

22.6.3.1.2 Backward Integration: This again is a vertical integration that moves backward. In this case, the farmer or integrator, in order to obtain his inputs at a cheaper rate, say chicks, he will start his own hatchery and / or breeding farm or a broiler retailer will start his own broiler farm to get greater profits by selling his own birds.

22.5.3.2 Horizontal Integration

It is the association, amalgamation or merging of the companies and units engaged in the same type of production, like merging together of two or more hatcheries, commercial farms, processing plants etc., resulting in increase in the volume of operation and expansion in production. The integrator can go for diversified fast foods and advertisement in mass media for sales promotion. Sometimes it may also lead to either unhealthy competition among integrators or even monopoly. Both are not good for the industry.

22.5.3.3 Parallel Integration

It is another type of integration, also called ad diversification. Here the integrator may start allied industries. For example, broiler farmer or breeder may start a feed mill to reduce feed cost as well as to sell feed. Similarly, a hatchery operator may go for manufacturing the incubators and use it for his own hatchery and sales. This type of integration, not only reduces the cost of production, but also obtain additional revenue, by selling the diversified products.

22.5.4 Advantages of Integration

1. Uncontrolled growth of independent small farms and other production units leading to congested overpopulated poultry pockets can be avoided.
2. The cost of production is less benefiting the consumer.
3. Sale of diversified products and newer fast foods are made available to the consumers as a marketing strategy of the integrators.

4. As the integrating companies involved in large volume of operation, they promote the consumption through advertisement
5. All main and by products can be utilized and recycled without any wastage. This will not only prevent environmental pollution; but also reduces the cost of production of the main product.
6. Integration stabilizes the price by balancing supply and demand.

22.6 Marketing Agencies

The wholesale market of eggs in larger cities are handled by few traders who have monopolized the market for their own advantage. As a result, neither the poultry farmer nor the consumer gets eggs at a reasonable price. This big margin of profit is retained by few traders who are exploiting the market. Thus, Small and large poultry co-operative societies have been set up in the country for marketing of eggs.

22.6.1 National Agricultural Co-operative Marketing Federation of India (NAFED)

National Agricultural Co-operative Marketing Federation of India (NAFED) has taken up egg marketing in New Delhi and extended to terminal markets like Mumbai, Chennai, Calcutta, Hyderabad and some other important industrial areas.

Main Objectives of NAFED

1. Provision of reasonable remuneration to the poultry farmers
2. Economic price to the consumer
3. Employment to the weaker section of the society.

22.6.2 National Egg Coordination Committee (NECC)

Around the year 1981, the Indian poultry industry was hit by an unprecedented crisis. Over 40 per cent of all poultry farmers had stopped operations because the business had become economically unviable. Middlemen had forced down prices and farmers were being paid less than their production cost, a result of speculative trading, since the existing market and distribution network was working against the interest of the farmers. Feed costs had risen by 250 percent in the past 5 years, whereas egg prices were static at an average of 35 paise. Consumption of eggs was low and the future looked anything but healthy. With no help coming from any quarter, a group of farmers motivated by Dr.B.V.Rao traveled across the country, organizing over 300 meetings with groups, individuals and traders. Their objective – unite poultry farmers from all over

India, and take control of their own destiny. Dr.Rao's call "My Egg, My Price, My Life" consequently brought farmers onto a united platform and realized this objective. NECC was formally registered under the Societies Registration Act in May 1982. NECC is unique in many ways. With a membership of more than 25,000 it is the largest single association of poultry farmers in the world. Most of today's egg production in India comes from NECC members. NECC has various centres and they exchange information among themselves which help them to determine price for each zone. Co-ordination between zones permits judicious movement of eggs dependent upon supply and demand.

NECC Objectives

1. Price declaration
2. To decide upon a reasonable price for eggs this ensures a fair return to the farmers, decent margins to the middleman and a fair price to the customer.
3. To monitor the egg stock level in different production centres.
4. To organize and unite poultry farmers across the country.
5. To generate employment by encouraging people to take up egg farming and egg trading.
6. To promote exports and develop export markets.

22.6.3 Farmer Producer Organizations (FPOs)

A Farmer Producer Organization (FPO) is a legal entity formed by primary producers, viz. farmers, milk producers, fishermen, weavers, rural artisans, craftsmen. A FPO can be a producer company, a cooperative society or any other legal form which provides for sharing of profits/benefits among the members.

The main aim of FPO is to ensure better income for the producers through an organization of their own. Small producers do not have the volume individually (both inputs and produce) to get the benefit of economies of scale. Besides, in agricultural marketing, there is a long chain of intermediaries who very often work non-transparently leading to the situation where the producer receives only a small part of the value that the ultimate consumer pays. Through aggregation, the primary producers can avail the benefit of economies of scale. They will also have better bargaining power vis-à-vis the bulk buyers of produce and bulk suppliers of inputs.

What are the Important Activities of a FPO?

The primary producers have skill and expertise in producing. However, they generally need support for marketing of what they produce. The FPO will basically bridge this gap. The FPO will take over the responsibility of any one or more activities in the value chain of the produce right from procurement of raw material to delivery of the final product at the ultimate consumers' doorstep. In brief, the FPO could undertake the following activities:

1. Procurement of inputs
2. Disseminating market information
3. Dissemination of technology and innovations
4. Facilitating finance for inputs
5. Aggregation and storage of produce
6. Primary processing like drying, cleaning and grading
7. Brand building, Packaging, Labeling and Standardization
8. Quality control
9. Marketing to institutional buyers
10. Participation in commodity exchanges
11. Export

22.7 Promotional and Advertising Techniques to be Followed to Improve Marketing Strategies

1. **Advertising:** Advertising includes various forms of paid media such as television, radio, print, and online advertising.
2. **Public Relations:** Public relations (PR) refers to the practice of managing the spread of information between an organization and the public. This can include press releases and media events
3. **Sales Promotion:** Sales promotion refers to short-term incentives to encourage the purchase of a product or service. This can include coupons, discounts, and contests.
4. **Direct Marketing:** Direct marketing is the process of reaching customers directly through mail, email, or telemarketing.

7. **Trade shows and events:** Trade shows and events provide an opportunity for companies to showcase their products and services to a targeted audience. Ex. Pashu Melas etc
8. **Digital Marketing:**Digital marketing includes online advertising, social media, email marketing, and search engine optimization (SEO).
9. **Influencer Marketing:** Influencer marketing refers to the use of social media personalities to promote a product or service.

10. **Referral Marketing:** Referral marketing refers to the process of encouraging customers to refer friends and family to a business.
11. **Affiliate Marketing:** Affiliate marketing is a performance-based marketing strategy in which a business rewards affiliates for driving traffic or sales to its website.
12. **Partnerships:** Partnerships refer to the collaborations between two or more companies to promote a product or service.

23

Specialty Eggs

Kiran M

Abstract

Meeting consumer demands is a constant challenge for the animal food industry. Many consumers desire somewhat distinct products with respect to safety, healthfulness, freshness, taste, color, etc. To tap into this market, companies have developed several designer and speciality eggs which have appeared on store shelves. The nomenclature of such eggs may be confusing, and often misleading. The purpose of this chapter is to discuss the types of new eggs currently on the market.

23.1 Introduction

Eggs are nutrient dense superfoods, naturally occurring vitamin pills. They contain a complete source of protein with all essential aminoacids and naturally occurring vitamin D. The adverse publicity of high cholesterol content in the egg limit their intake. Nowadays consumers need the products free from drugs/ pesticides and contain special health promoting substances. To reap the growing demand of health conscious consumers and to make available of their requirements, designer eggs are made. Designer eggs are those eggs their composition differ from the standard eggs by changing the feed of the hen. They are manipulated to contain more amount of vitamins, minerals and pigments and also to deliver the pharmaceutical compounds and biological compounds. This will help consumers which are all ailing from cardiac illness, diabetes, macular degeneration problems and many other health issues.

23.2 Designer Eggs

One of the ways to market a new product is to change the old product. The contents of the chicken egg can be changed in such ways as to be more healthful and appealing to a segment of our consumers who are willing to pay for those changes in the egg. Designer eggs are those in which the content has been modified from the standard egg.

23.2.1 Vitamin Content

Designer eggs have been produced that contain higher concentrations of several vitamins. Two vitamins, A and E, are receiving the most interest as components of designer eggs. The vitamin content of the egg is variable and is somewhat dependent on the dietary concentration of any specific vitamin. In addition, the hen does not transfer different vitamins into the egg with equal efficiency. Because of this, the vitamin transfer efficiency and cost of the vitamin must be taken into consideration when determining the economic feasibility of marketing such eggs. Eggs higher in Vitamin E are currently available in stores.

23.2.2 Fat and Fatty Acid Content

Altering the total fat content in the diet of the hen has little effect on the total fat content of the egg yolk. However, the fatty acid profile (or the ratios of the different types of fatty acids) of egg yolk lipid can easily be changed, simply by changing the type of fat used in the diet. Consumption of polyunsaturated fatty acids has been reported to reduce the risk of atherosclerosis and stroke. Consumption of these fatty acids has also been shown to promote infant growth. Different feeds, such as flaxseed (linseed), safflower oil, perilla oils, chia, marine algae, fish, fish oil, and vegetable oil have been added to chicken feeds to increase the omega-3 fatty acid content in the egg yolk. Omega-3 fatty acid-rich eggs may provide an alternative food source for enhancing consumer intake of these 'healthy' fatty acids. Evaluation of the eggs during storage indicates that the shelf life of the enriched eggs was comparable to that of typical eggs. Omega-3 fatty acid-enriched eggs taste and cook like any other chicken eggs available in the grocery store. However, they typically have a darker yellow yolk. There are also designer eggs on the market that contain a lowered saturated to unsaturated fatty acid ratio. Canola oil is commonly used to alter the ratio of saturated to unsaturated fatty acids.

23.2.3 Cholesterol Content

Even though the dietary cholesterol is insignificantly correlated with the serum cholesterol levels, the consumers are scared of high cholesterol foods, like eggs. A large egg contains about 200 mg of cholesterol. Research towards lowering egg cholesterol has centered mostly on dietary and pharmacological interventions. Chromium, copper, nicotinic acid, statins, garlic, basil (tulasi), plant sterols, N-3 PUFA supplementation to chicken feed will reduce the yolk cholesterol levels significantly. Similarly, dietary linseed oil, fish oil, garlic, basil, spirulina, bay leaves, nicotinic acid, neomycin, yeast, guar gum, grape seed pulp / tomato pomace (lycopene), citrus pulp (nirangenin), chelated copper, organic chromium, roselle seeds and many more herbs in chicken diets will reduce the yolk and chicken fat cholesterol levels by 10-25%.

23.2.4 Mineral Content

The shell contains the majority of the minerals in an egg. There are approximately 2,200 mg of calcium and 20 mg of phosphorus in the shell. There has been very little success in changing the calcium and phosphorus content of the albumen and yolk. It is possible, however, to increase the content of selenium, iodine and chromium. This has been done through dietary supplementation of the hen. These three minerals are important in human health. There has been some interest, therefore, in promoting these eggs as designer eggs.

23.2.5 Pigment Content

The color of the yolk is a reflection of its pigment content. In addition, the type of pigment in the egg and its concentration are directly influenced by the dietary concentration of any particular pigment. Consumer preferences vary greatly on yolk color, even in the same country. Color is described on the basis of the Roche Color Fan (RCF). Yolk colors from 6 to 15 can be achieved by using only natural pigmenters obtained from natural raw materials. Natural sources can be from plants such as marigold, chili, or corn. The high protein blue-green algae known as Spirulina has also been shown to be a very efficient pigment source for poultry skin and egg yolk. Eggs, due to the presence of large quantity of pigments are beneficial in preventing macular degeneration, a major cause of blindness in the elderly. The most effective carotenoids were lutein and zeaxanthin, which are commonly found in dark-green leafy vegetables, such as spinach and collard greens. Most of the carotenoids in egg yolk are hydroxy compounds called xanthophylls. Lutein and zeaxanthin are two of the most common xanthophylls found in egg yolk. Lutein and zeaxanthin are high in pigmented feed ingredients such as yellow corn, alfalfa meal, corn gluten meal, dried algae meal, and marigold-petal meal. Fortunately, both lutein and zeaxanthin are efficiently transferred to the yolk when these various feed ingredients are fed to laying hens. The egg processing industry has routinely produced highly pigmented yolks for use in bakery products, pasta and mayonnaise. Perhaps there is a market for eggs having a higher level of lutein and zeaxanthin. With a growing problem of macular degeneration in the elderly, the egg industry may can seize this opportunity.

23.2.6 Pharmaceuticals

New biotechnology is being used to develop genetically modified chickens that produce compounds that can be harvested from the eggs. These compounds include insulin for the treatment of diabetes. The hen, like all animals, produces antibodies to neutralize the antigens (viruses, bacteria, etc.) to which she is exposed to each day. These antibodies circulate throughout her body and are transferred to her egg as protection to the developing chick. Immunologists are taking advantage of the fact that the hen can develop antibodies against a large array of antigens and concentrate them in the egg. Specific antigens are now being selected and injected into the hen who develops antibodies against them. As new biotechnology knowledge is gained in this area, designer eggs in the future may be produced that result in a range of antibodies for treatment against snake venoms, to the countering of microorganisms which cause tooth decay.

23.3 Specialty Eggs

Eggs which have a special attribute which makes them attractive to a niche market are known as specialty eggs. A number of these eggs are available in stores.

23.3.1 Cage-free or Free-roaming Eggs

The majority of commercial egg layers are housed in cages. Caging of hens has benefits for the birds, consumers and producers. The separation of birds from their faeces is advantageous in reducing the risk of disease and parasitic infections. Working conditions for producers are often better with cages than with other systems: automation is possible reducing the amount of physical labor, and dust and ammonia are usually less prevalent. Eggs are laid on the sloping floor of the cage so that there is minimal contact between the egg and the hen. This decreases the possibility of bacterial contamination of the egg.

With regard to the welfare of the laying hens, cages have both advantages and disadvantages. While the issue of cages and animal welfare is still being debated, there are some consumers who prefer to purchase eggs produced from hens that are not kept in cages. In order to meet this niche market, some producers raise their birds in a cage-free or free-roaming system. It is important to note that cage-free does not mean that the birds are raised outdoors. Typically the birds are maintained on the floor of a poultry house. The higher production costs associated with this type of management system are reflected in the higher price for the eggs. The price of cage-free eggs is often twice that of regular eggs.

23.3.2 Free-range Eggs

Free-range eggs are produced from hens that are allowed to graze or roam outdoors. It is not necessary, however, for the hens to be outdoors all the time. Typically, the hens are housed in a poultry house that has access to the outdoors. The hens have the ability to go outdoors during the day, although they can also choose to stay indoors. The flock is usually locked indoors at night to protect the hens from predators. There is no set standard on how much range must be available for the hens.

23.3.3 Pasture Rised Eggs

Pasture rearing of chickens is a modification of the free-range system. The birds remain on pasture all the time, but are confined within a portable pen. The pen is moved daily to give the birds access to fresh pasture. The portable pen usually has a portion covered to protect the hens from the elements.

23.3.4 Organic Eggs

Until recently there was no set standard for the production of organic poultry products. The USDA is currently working on developing legal standards. Many states of USA, including Florida, have set standards for organic produce. To be certified organic, the eggs must be produced from hens that have been fed certified-organic feed which was produced without synthetic pesticides or herbicides, antibiotics, or genetically-modified crops. In addition, no synthetic pesticides can be used to control external and internal parasites. Typically organic eggs are also produced from hens in cage-free systems.

23.3.5 Fertile Eggs

Almost all eggs produced commercially are infertile. Roosters do not have to be present for hens to lay eggs and roosters are, therefore, not kept with laying flocks. When there is an excess of hatching eggs in the poultry meat industry, eggs from broiler breeder flocks can be sold for human consumption. A large percentageage of these eggs will be fertile. Fertilized eggs are safe to eat. There is no nutritional difference between fertilized and unfertilized eggs. The embryo does not develop in fertilized eggs that are refrigerated soon after laying.

23.7.1 Production of Low Cholesterol Eggs

If the egg yolk cholesterol content could be reduced to 50%, the consumption of two eggs per day will meet the AHA. Efforts such as genetic selection, nutritional alternative and anticholesterol drug addition have been pursued extensively to reduce egg cholesterol levels.

23.4.1.1 Genetic Selection

From a genetic point of view, egg cholesterol content is affected by several factors, such as species of bird, breed or strain and age of fowl. In the 1970s several genetic selection studies aiming to decrease egg yolk cholesterol were carried out. Reduction in egg cholesterol by genetic selection is fairly small (9 to 10 mg) and is not significant if an average daily intake of cholesterol is about 250 mg. However, there is a lower limit to egg yolk cholesterol, due to its role in developing the embryo. Another disadvantage is that genetic selection for low cholesterol eggs results in smaller yolks and smaller egg size. However limited changes in yolk cholesterol concentration (5 to 7% reduction) can be achieved by genetic selection.

23.4.1.2 Diet

Effect of fat: Dietary cholesterol supplementation can increase the concentration of cholesterol in eggs. Dietary omega-3 fatty acid is another factor affecting yolk cholesterol content. Yolk cholesterol content in omega-3–enriched eggs has less yolk cholesterol. Feeding of birds with menhaden fish oil and flaxseed results in yolk cholesterol reduction.

Effect of Natural Products: Some natural products have been shown to have the ability to reduce plasma cholesterol level in humans and animals. Garlic has been reported to be the most effective agent in reducing the cholesterol level in yolk and plasma.

Effect of Phytosterols: Phytosterols (also called plant sterols), which are structurally and functionally similar to cholesterol in vertebrate animals. Feeding plant sterols decrease cholesterol concentration in plasma and egg.

Effect of Minerals and Vitamins: Dietary microminerals (copper, zinc, vanadium, chromium, and iodine) and/or dietary vitamins (vitamin A, ascorbic acid, and niacin) change the yolk cholesterol level. However, the role of vitamins and minerals has not been proved yet beyond doubt.

Effect of Fibre: Dietary fibre is a group of indigestible portion of plant foods that can be resistant to the human or animal gastrointestinal system, absorbing water and easing defecation. It is thought that fibres could affect cholesterol metabolism through bile acid binding and stimulate faecal sterol excretion. Regardless of the dietary source, feeding fibre to laying hens dilutes the available energy content of a diet and, as a result, may limit energy intake and potentially reduce hepatic cholesterol production, especially if prior energy intake had been excessive.

23.4.1.3 Pharmacological Methods

Certain drugs have been successful in lowering egg cholesterol by as much as 50%. Drugs lower cholesterol in the egg by either inhibiting the synthesis of cholesterol in the hen or by inhibiting the transfer of cholesterol from the blood to the developing yolk on the ovary. Today, the drugs which have shown promise in lowering cholesterol are not yet approved by the FDA for commercial use. Chromium supplementation to laying hen diets at concentrations of less than 1 ppm have been shown to lower egg cholesterol and also improve egg interior quality. Even though genetic selection, alternative nutrition, and pharmacological techniques are an effective strategy in modifying egg cholesterol content, there is a certain limit possible for the reduction of cholesterol levels. Research needs to be done in the avian transgenesis field, which seems to be the most effective way to reduce the cholesterol level in eggs. Furthermore, the commercial ability to produce low-cholesterol eggs will be constrained by environmental policies, biotechnology, and public acceptance.

11.4.1.3 Pharmacological Methods

24

Value Added Convenience Egg Products

Kiran M., Wilfred Ruban S. and Deepika J.

Abstract

Egg products are quick, hygienic, convenient, modern and 100% natural products. They represent a new way to consume eggs meeting the requirements of contemporary consumption. Once processed, eggs are easier to preserve, transport, store and use. Not only are they practical, safe and healthy products, but they are also similar to raw eggs in flavor and nutritional values. Growing industrialization of the egg-based products, tightening of health standards related to food safety or evolution of the final customers' ways of consuming eggs in a processed form.

24.1 Introduction

Egg Products have gained great acceptance all over the world to-day as they offer advantage in labour saving and time in preparations, space saving and consistency in quality. There is also an an increase in consumer demand for a variety of preparations to suit their taste and purse. Today's commercial bakeries, food processing plants and institutions use tremendous quantities of liquid bulk eggs, egg solids (dried egg) and frozen egg products for preparation of several convenient foods. Therefore food technologists all over the world are attempting to provide a variety of value added/convenient egg products to meet the demand of the consumers. Several such preparations have come up in India as well as in many other countries. Some of the preparations from eggs have pharmaceutical and industrial applications too.

24.2 Common Egg Products

24.2.1 Boiled Eggs

Depending on the consistency of cooked egg, it can be called as soft-cooked or hard-cooked egg. For cooking, the eggs stored in refrigerator should be taken out at least half an hour before. Eggs that are too fresh, that is, less than five days old are difficult to peel. Therefore, older eggs would be ideal for boiling. The next point to consider is the size of the pan. Small pan, or a pan that is just large enough for the number of eggs to be boiled. This prevents the eggs from breaking, as the eggs in larger pan hit the other eggs or the sides of the pan during cooking. The pan with enough water to a level of one inch above the eggs is added with a teaspoonful of salt. This will make peeling the egg easier once it has been cooked. Bring the water to the boil. Once water is boiled, pierce the larger, rounder end of the eggs with a pin and carefully lower the eggs into the pan using a tablespoon. As the egg heats up during cooking, the air cell within the egg swells and can cause the egg to crack due to the pressure and lack of escape. The pinprick in the end of the egg will allow the steam to escape instead of cracking the shell. As the water reaches boiling point again, reduce the heat so that the water is simmering and begin timing. If eggs are cooked at temperatures that are too high, this will cause the egg white to toughen. The soft-cooked eggs can be prepared by keeping the eggs in boiling water for 4 – 6 minutes and hard-cooked eggs for 10 minutes. As soon as the cooking time is up, remove the eggs from the boiling water and run them under cold running water or immerse them in a bowl of very cold water. This will stop the eggs from cooking in their own heat and will prevent a discolouration of the egg yolk that sometimes forms. The eggs are rolled over a hard surface or gently tapped against a hard surface in order to crack the shell before peeling. Cooked unpeeled eggs can be stored in the refrigerator for a week period. It has to be remembered that the timings has to be adjusted depending upon the size of the eggs. The time required in ascending order is quail, followed by Guinea fowl, chicken, duck, turkey and goose. Hard-cooked eggs - can be served whole or they may be cut and sliced. An egg slicer can be used to prepare evenly sliced sections that can be used in salads or as a garnish.

24.2.2 Deviled Eggs

Hard-cooked eggs can also be cut in half the long way so that the solid yolk can be removed from the two halves. The yolk is then blended with other ingredients such as mayonnaise, mustard, and seasonings and the mixture is stuffed back into the yolk cavity of the egg halves and served.

24.2.3 Coddled Eggs

A coddled egg is cooked more slowly than a boiled egg, but basically yields the same results, except that the egg is a bit tender. Pierce the large end of the eggs with a pin. Place the eggs in a single layer and add cold water and 1½ teaspoons of salt in the saucepan and leave the pan uncovered. Place the pan on medium heat and bring the water to a simmer, but not to a full boil. Remove the pan from the heat and cover it. The length of time that the eggs remain in the covered pan determines the degree of firmness of the yolk: soft yolk for 4 to 6 minutes, medium yolk for 6 to 8 minutes and hard yolk for 20 to 25 minutes. To stop the cooking process, run cold water over the eggs. It is best to use older eggs for coddling because they peel easier. Soft-cooked coddled eggs are often served in an egg cup and eaten directly from the shell because they are difficult to peel.

24.2.4 Roasted Eggs

A roasted egg is prepared using two different cooking processes. It is first hard-cooked in simmering water and then it is placed in the oven and roasted in the shell. It is removed from the oven when the shell becomes brown. This method is used to prepare eggs traditionally served during the Jewish Passover. An important point to consider in order to achieve good results (regardless of the cooking method) is that eggs should not be overcooked. Eggs cooked too long or at too high a temperature may become tough and rubbery and will not be very appealing. Unless eggs are rapidly cooled, they continue cooking after removal from the heat source.

24.2.5 Omlette

An omelet is usually made with 2 or 3 eggs and is cooked very quickly in a sauté pan. Salt is then rubbed on the pan and wiped clean after each use. The salt treatment helps prevent eggs from sticking to the pan the next time it is used. Additional ingredients, such as curry leaves and other herbs, cheese or finely chopped meats can be added to the eggs before they are cooked without changing the way the basic omelet is cooked. The extra ingredients will probably require that a spatula be used to fold the omelet in the pan rather than trying to roll it over by jerking the pan. The extra ingredients can also be placed on top of the plain omelet when it is served.

24.2.6 Scrambled Eggs

To prepare scrambled duck eggs, break the eggs into a bowl, add a tablespoon of cold water, and whisk together so that the yolk and whites are blended. Salt and pepper can be added to the eggs or added when the eggs are served. Coat the pan lightly with oil or butter, heat it over a medium heat, and pour the egg mixture into the pan. As the mixture begins to set, use a spatula to scrape the eggs from the edge of the pan to the center. It is best to remove the eggs from the heat source when they are still very moist, because the internal heat of the eggs will complete the cooking process. When 2 or 3 eggs are scrambled together, it usually takes a minute or less before the cooking process is complete. Eggs that become too dry indicate that they have been overcooked. For increasing the quality of duck egg scrambling, lemon extract may be helpful. The citric acid content of lemon will adhere the ovomucin factor which will effectively reduce the duration of scrambling. The duck egg has a stronger flavor than a chicken's egg. Therefore, duck eggs either in scrambling or in omelets are well complemented by onions, peppers, mushrooms, or cheese.

24.2.7 Poached Eggs

A poached egg cooked on the stovetop is one that is cooked in simmering water without the shell. Unlike a boiled or coddled egg that benefits from the use of an older egg, a poached egg is best when a very fresh egg is used. This is because the fresh egg, when placed into the heated water, will not spread out like an older egg, yielding better results with the shape and texture of the egg. If an older egg must be used, it can be simmered in the shell for a few seconds so that the white is just slightly coagulated. When the egg is broken into the simmering water, it will not spread out as much. One tablespoon of vinegar added to the water will also help with coagulating the white to keep it from spreading too much. In another method, an oval, slotted metal container (poacher) with an attached handle is used. The metal poacher is convenient to use and creates a pleasing shape. Both fresh eggs and older eggs can be poached well using the metal poacher. If the egg is to be poached in a microwave, there are various poaching dishes that can be used, such as the cooking dish.

24.2.8 Balut

Consumption of fertilized duck eggs are familiar in the food customs of Chinese, Laotians, Cambodians and Thais. This is a duck egg that has been incubated 17 days and removed for boiling and consumption. It is considered a delicacy and is highly nutritious.

24.2.9 Salted Duck Eggs (Uncooked)

The salted duck eggs are famous for many centuries. They are characterized as being loose, oily, fresh, fine, tender, and tasty. They salted by traditional Chinese method. The end product is having mild salted flavour and gold yellow yolk.

24.2.10 Custards

In custards egg acts as thickening agent. True custard consists only of eggs, milk, sugar and flavoring custards are of two types, the stirred or soft custard.

24.2.11 Souffle

They are similar to foamy omelette, except that have a thick white sauce base and contain additional ingredients such as grated cheese, vegetable pulp or ground meats. Dessert souffles are sweet and may contain lemon, strawberry and chocolate. Souffles are usually baked though they can be steamed. Crepes are thin, tender pan cakes containing a relatively high proportion of egg. They can be filled with a variety of items, including fish, meat, poultry, cheese, vegetables or fruits. Crepes can be served with sweet, dessert type fillings.

24.2.12 Egg Pickles

Pickled eggs are prepared from top grade hard cooked eggs. Most pickled eggs are small size eggs, about 40-45 grams in weight including shells. Preparation through the cooling and peeling operations is the same as described for the hard cooked egg. The pickling solution used varies widely among producers. The basic ingredient is vinegar or 5% acetic acid (food grade) solution. Other possible ingredients are red cinnamon candy, pickling spice mix, salt (sodium chloride), garlic or garlic salt, fresh sliced onion, mustard seed, sugar and red beet juice. The simplest pickling solution would contain only vinegar, salt and pickling spices. For faster penetration, the pickling marinade is usually poured over eggs as a hot solution.

24.2.13 Apple Egg Drink

This was developed in Cornel-USA combining various fruit juices with egg. Since New York is a large apple producing state in USA, where a drink made of apple juice and egg was taken up for market testing. The name chosen for the product was TREN-a combination of tree and hen. It was known as a breakfast- in- glass since the egg contains all the nutrients except Vitamin-C and this can be easily added to the egg. The ingredients included apple juice, fresh egg, sugar, citric acid and apple essence.

24.2.14 Flavoured Eggs

This is a value added egg product most common in China. In this product , spices are added to the egg soaking solutions to create so called flavoured Eggs. Composition includes 25% edible salt, 2% black tea, 5-10 % (such as clove, spices mixture, anise, pepper or combinations) with immersion at 25°c for a period of 20 days. Eggs are then taken out and boiled for 5 minutes before serving or further smoked to produce smoked, flavored eggs Tea flavoured eggs are also popular in china. The difference between flavoured and tea flavoured eggs is that the eggs for tea flavouring are first boiled and cooked and the egg shell is broken slightly. The eggs are then slightly immersed in a solution of salt, tea leaves, and spices and cooked for few hours to create the unique flavour.

24.2.15 Sour Eggs

For preparing sour eggs, eggs are immersed in a solution of vinegar and edible salt until egg shell is partially dissolved or softened. The egg is then immersed in wine distillate with edible salt and a small amount of vinegar. Ageing is completed after 3-5 months of soaking. The sour egg is further heated to 80°c for 5 minutes before serving. Sour egg is a product of acid coagulation and it contains the unique sour and wine tastes.

24.2.16 Egg Tofu

This is a traditional East Asian food, which is made from whole egg, mixed with water; salt, monosodium glutamate, sucrose, soy sauce and fish protein extract and then heated to form an egg gel. Its texture is identical to that of bean-curd (Tofu), hence the name. Production of tofu is on the principle of heat coagulation of egg albumin and egg yolk proteins.

24.2.17 Scotch Eggs

The Scotch egg is a popular snack food. It is commercially manufactured and widely distributed in pubs, railway stations snack bars, delicatessens, restaurants. The Scotch egg, gourmet item consists of hard cooked, peeled egg wrapped in a layer of sausage and then deep-fat fried.

24.2.18 Egg Salads

Egg salads vary widely in total composition. Two essential ingredients in egg salads are salad dressing and chopped hard cooked eggs. Many egg salads also have pickle relish and chopped onions.

24.2.19 Long Eggs

A variation of hard cooked eggs is the long egg. This is prepared by cooking a uniform layer of albumen around a centre core of egg yolk in a long cylinder shape. Each slice of hard cooked long egg looks like the centre cut of a normal hard cooked egg.

24.2.20 Crepes

A popular egg-rich product is crepes. They are prepared as a very thin pancake batter particularly high in egg content. So thin crepe will maintain integrity when folded. Crepes are cooked and then may be filled with vegetables, meat, sea food, fruits are cream pie fillings. Crepes to be filled should be circular in shape and about 8 inches in diameter. Immediately after cooking they are placed on the filling line where the desired amount of filling is placed near the centre of each cream. It is then rolled by hand and placed in a suitable package. Some crepes are folded rather over the filling.

24.2.21 Pancakes

Frozen pan cakes are a convenient item for the busy home maker. To achieve a light, fluffy pan cake, the batter should be rich in eggs and extra whipping to yield a foamy structure. Frozen pancakes are cooked in a manner similar to that for crepes but longer cooking time is required as the pan cakes are thicker than the crepes. The amount of batter measured for each pan cake is carefully controlled to get a pan cake of uniform diameter of about 5 inches.

24.2.22 Waffles

Waffles are easy to prepare using a micro wave range, a toaster or a conventional oven. The waffle batters for the best quality waffle should be high in eggs for a light, tender waffle. A waffle batter formula is very similar to that used for pan cakes.

24.2.23 Eggurt

Eggurt is a yogurt type product containing egg albumen as a partial milk replacer. It is prepared unto 45% of the milk replaced with albumen.

24.2.24 Egg Rings

Egg rings are prepared by cooking albumen in ring molds, battering and breading the coagulated albumen, and deep fat browning of the breading. They look much like battered fried onion rings.

24.2.25 French Toast

It is prepared by soaking slices of bread in a blended liquid egg mixture similar to the formula for omelets and then pans frying the coated bread. French toast is made from special bread which has a multitude of small pores. The bread is frequently yellow which helps give the finished product a uniform color, even when liquid egg does not penetrate to the centre of each slice. If the excess egg is use the cost of production increases, but more importantly the toast does not cook through to the centre and may be soggy.

24.2.26 Egg Rolls

It consists of baked bread with scrambled eggs and sausage filling that is rolled like a jelly cake prior to packing. Other items such as onion, herbs, spices, cheese bacon and canned meat might also be included. To prepare egg roll-ups bread dough is rolled out to about 1 cm thickness in a rectangular shape. Scrambled eggs are cooked to a soft done condition. Other ingredients are also cooked and are layered on the bread along with eggs. The layered ingredients must be kept about from 1 cm from the edges. The product is then rolled and backed.

25

Organic Eggs

Vidyasagar

Abstract

FAO defines Organic farming as " a unique production management system which promotes and enhances agro-ecosystem health, including biodiversity, biological cycles and soil biological activity, and this is accomplished by using on farm agronomic, biological and mechanical methods in exclusion of all synthetic off farm inputs."

25.1 Introduction

There has been a recent increase in the consumption of organic food, particularly in developed countries. Consumer awareness is growing in terms of organic food products in recent years as almost all the food ingredients are grown under intense production systems which utilize lot of chemicals and pesticides to control the pests and diseases. Organic farming can be defined as an approach to agriculture where the aim is to create integrated, humane, environmentally and economically sustainable agricultural production systems producing acceptable levels of crop, livestock and human nutrition, protection from pests and diseases, and an appropriate return to the human and other resources employed. Maximum reliance is placed on locally or farm-derived, renewable resources and the management of self-regulating ecological and biological processes and interactions. The usage of chemical and other external inputs are reduced as far as possible. Organic agriculture is known as ecological agriculture, reflecting this reliance on ecosystem management rather than external inputs. In India free range farming or backyard poultry farming is considered to be organic if birds are reared without any medication and other feed compounds.

25.2 Organic Poultry Farming in India

India exported organic agriculture of product worth Rs. 72 crores during 2004-05, but almost all products exported from India were of plant origin, whereas India has large number of livestock and poultry population. Even if a small shift from current conventional production to organic animal production can create a huge market to domestic consumption as well as export well as export Our country has a vast scope for promotion of organic farming in the export market, without compromising with the national food security as farming by tribals and under rainfed conditions is generally organic, since very little chemical intput are used.

25.3 Backyard Poultry Farming

Backyard or homestead poultry farming is common among rural and landless families in India and is a lucrative source of supplementary income. It involves low investment and yields high economic returns, and can be easily managed by women, children and the elderly. Meat and eggs from such birds are inexpensive and rich source of protein and energy for poor households. Backyard poultry farming is characterized by an indigenous night shelter system, scavenging, natural hatching of chicks, low productivity of birds, scant supplementary feed, local marketing and minimal health care practices. In this system, flock size ranges from five to 50 birds raised under a traditional scavenging system devoid of management practices. Backyard chickens in India are generally the native/desi type (19 poultry breeds) with low egg and meat production potential. Desi chicken breeds grown under free range backyard conditions contribute about 10-12% of the total egg production in India.

25.4 Difference Between organic and Free Range / Backyard Poultry Farming

S.No.	Organic Poultry Farming	Free Range Poultry farming
1.	Organic hens are fed only organic feed and it is prohibited to feed animal by products or GMO(Genetically modified crops)	Birds are left outside environment and birds are allowed for scavenging where the feeding is not prohibited
2.	No antibiotics allowed except in emergencies	In free range, it is up to the farmer, but the same levels of antibiotics as conventional farming is allowed
3.	Required animal welfare standards in organic farms are higher, which can improve the quality of both the eggs and the meat.	Standards are not upto the level

25.5 Basic Requirements for Organic Poultry

1. Appropriate housing that permits natural behaviour, including outdoor access
2. Certified organic feed should be provided
3. No antibiotics, drugs or synthetic parasiticides
4. Organic processing of meat and eggs
5. Recordkeeping system to allow tracking of poultry and products (audit trail)
6. Organic system plan including description of practices to prevent contamination, monitoring practices and list of inputs
7. Production that does not contribute to contamination of soil or water
8. No genetically modified organisms, ionizing radiation or sewage sludge
9. Stress management
10. Access to the environment.
11. Access to the pasture.
12. Natural maintenance of shelter.
13. Avoidance of synthetic substances.
14. Rearing of birds without cages
15. Natural treatments

25.5.1 Breeding

The choice of breeds of poultry should be based on their adaptability to the local conditions. Breeding goals should not be contrary to the animal behaviour and should be directed towards good health. The use of genetically engineered species or breeds is not allowed for organic farming. Reproductive techniques should be natural and hormonal treatment for better egg production is contraindicated. The sex ratio should be one male for about 4-6 hens in a flock like in case of the wild birds.

25.5.2 Housing

The birds should be housed in such a way as to provide opportunity for the bird to exhibit all its normal behaviour patterns and experience minimal stress. Cages should be avoided and birds should be reared under deep litter system. They must have easy access to an outside grazing area, fresh air, clean water, balanced ration, dust bathing facilities and an area for scratching. Debeaking and beak trimming are usually prohibited practices but some certifying agencies still permit Debeaking if done more than 5mm of the upper beak should not be removed In the organic meat sector birds must be grown for a period of 81 days of age usually.

25.5.3 Feeding

Birds should be fed 100 per cent organically grown feed of good quality. All ingredients must be certified as organic, except vitamin and mineral supplements making up to 5 per cent of the diet. The diet should be offered in a form so that the birds can exhibit natural feeding behaviour and digestive needs. Concentrated balanced feed ration produced organically should be given. The largest component of any organic poultry diet is maize and high quality roughages, particularly legumes can be supplemented to the diet. Home grown protein sources like peas, beans and rapeseed can also be utilized. Peas can be included at the rate of 250-300g/kg for the table chicken and 150-200g/kg for laying hens. Since sprouted pulses are a good source of vitamins they can be preferentially used to replace synthetic amino acids. Limestone and rock phosphate in general and limestone grit and rock phosphate particularly for layers can be included as mineral source. Trace minerals incorporated in the diets should be organic or ayurvedic in nature. The quota of essential amino acids can be met through feeding organic soybean, skim milk powder, potato protein, maize gluten etc. Nevertheless, balanced ration should be given and over feeding must be avoided. A continuous access and ample supply of drinking standard quality water free from residues should be assured. Water should be regularly tested for ground water contamination.

Feed used for organic poultry production must not contain

1. Animal drugs, including hormones to promote growth.
2. Feed supplements or additives in amounts above those needed for adequate nutrition and health maintenance.
3. Plastic feed pellets.
4. Urea or manure.
5. Mammalian or poultry slaughter by products fed to mammals or poultry.
6. Feed, additives, or supplements in violation of the Food and Drug Administration.
7. Feed or forage to which any antibiotic, including ionophores, has been added.

25.5.4 Treatment

Vaccines and Probiotics are allowed. Antibiotics treatment is must not be with held if birds need these birds are diverted in to non organic markets.

25.5.5 Predator & Rodent Control

Poultry should be protected from predators both indoors and outdoors. Electric fences for outdoor and for indoor rodents habitat reduction should be done by physical exclusion or mechanical and physical methods like traps, electric fences adhesives and fans can be used.

25.5.6 Health Care

The principle of healthcare and management in the organic farming concept is that when all management practices are directed towards the well being of the birds they will achieve maximum resistance against diseases and overcome many infections. Clean grazing and dry litter would ensure prevention of almost all health related problems. Use of antibiotics should be avoided; however vaccinations are permitted only when diseases are expected to be a problem. Use of natural medicines and homeopathy and ayurveda should be encouraged.

25.5.7 Record Keeping

Record keeping with respect to the overall management practices is the most important factor. It should be systematic documentation of activities, observations and inferences from time to time for future reference. Records include breeding records, registers indicating source of animals purchase, source of organic feed ingredients, feed supplements and feed additives purchased, organic feed formulation record, organic poultry pasture record, inventory of health care products, sanitation products, monthly flock records of organic egg layers, organic meat poultry, organic poultry slaughter/sales summary and monthly organic egg packing /sales record.

25.6 Regulations for Organic Poultry Farming

Organic poultry production in India at the moment is not regulated by any formal standards at national level except few prescribed standards by international agencies. All Indian producers who want to have their products labelled as organic poultry must in effect comply with the international / IFOAM standards. In an international context, the IFOAM standards for organic livestock production underpin most national organic livestock standards which are not otherwise covered by legislation, and these standards have had some impact on the drafting of international trade agreements such as the FAO Codex Alimentarius definitions and WTO agreements. Further, the IFOAM standards do not specify much detail relating to poultry production, but deal more with general principles. However, a more critical review of some of these standards with respect to organic poultry for India is necessary. Developing the processing and marketing standards for poultry, including the optional use of indications concerning the type of farming (specifically: extensive indoor (barn-reared), free-range, traditional free-range and free range: total freedom) is need of the hour.

25.7 Constraints for Organic Poultry Farming in India

1. Lack of in-depth knowledge about organic poultry farming on the part of poultry farmer and awareness among consumers is a hurdle in both at the production and marketing level.
2. Inadequate Supporting Infrastructure like lack of adequate financial support, inadequate local certifying agencies and lack of marketing channels
3. Strict measures mainly sanitary conditions, quality and traceability followed by developed countries is an obstacle for small and marginal Indian poultry farmers to enter into export of organic products.
4. Training facilities for poultry farmers are not adequate

In true sense 'livestock revolution' aims not merely increasing the quantum of production but to have an holistic approach to improve food security and safety of consumers. India has tremendous potential in organic poultry production as large part of country is organic by default. On the other hand the ill effects of conventional farming are compelling the consumers to shift to the organic products. So, The organic poultry is poised to transform poultry sector in particular and animal-agriculture in general, if regulations, infrastructure facilities, transfer of technology and sectoral cum target oriented development programmes are brought in practice with basic thrust towards 'food safety and poultry welfare'. What we need today is the necessary institutional and policy framework that can pave the way for the promotion of organic poultry farming in particular and organic livestock farming in general on a nationwide scale.

26

Non Food Uses of Eggs

Deepika Jamadar and Kiran M

Abstract

The egg is a staple food with a high nutritional value which also has several uses beyond food. More than a million eggs are processed each day in more than one hundred egg factories.Since the beginning of time, eggs have been utilized for purposes other than food, including artistic and cultural endeavors, experimental use, cultural media for a variety of microorganisms growth, and used in manufacturing many vaccines. Eggs play a significant role in the production of several kinds of cosmetics and shampoos, leather, and adhesives. They are also widely used in the production of fertilizers and are a vital component of animal feed.

26.1 Introduction

The egg's primary non-food purpose is to help the species reproduce. The quantity of eggs required for the hatching of chicks has increased significantly due to the substantial rise in meat consumption. The USDA has made a list of a few other non-food use for eggs.

Inedible eggs are often used as animal feed and plant fertilizers. Eggs are also fed to show animals, such as dogs and horses, to improve the glossy sheen of the fur coat. Industrial egg albumen is used in finishing several types of leather, notably coloured stock, and glazed. production of vaccines in chick embryos is one of the key uses of eggs, as per various researchers.Chick embryos are used to mass-produce vaccines for a wide range of infectious diseases. Applications in art and culture are very different ways to use eggs. Eggshell decorations were used as gifts for many ancient emperors, and eggshell fragments were utilized to create many varieties of mosaics. An egg yolk emulsion thinned with water is also used in tempera painting (dries quickly and stable).Eggs have been utilized throughout history in creation, fertility, purity, resurrection, and witchcraft. The egg has often been considered a sign of life, and some ancient cultures even prohibited the consumption of eggs. Membranes from eggshells are usually regarded as waste. Scientific studies in Japan, however, have demonstrated the value of protein products manufactured from hydrolyzed egg membranes in promoting wound healing and skin regeneration. Many cosmetics and shampoos today use egg membrane protein because of its emollient qualities. Red table wines are clarified using egg whites.

26.2 Experimental Uses

In general, eggs are considered as a rich source of high-quality protein. Eggalbumen or white has a very high biological value in terms of the quantity and balance of amino acids. In the lab, animals like the Chick, rat, mouse and others are used to compare proteins from numerous sources to egg white as a standard. The Food and Agriculture Organization's (1965) chemical score for the essential amino acids in protein was based on the amino acid profile of a complete hen egg. Nowadays, sodium dodecyl sulfate-polyacrylamide gel electrophoresis routinely uses individual egg white proteins as molecular weight standards. To produce protein markers with molecular weights of 14,600, 45,000, and 75,000 daltons, respectively, lysozyme, ovalbumin, and conalbumin are utilized.

26.3 Biological Uses

26.3.1 Culture Media

Egg yolk is an excellent medium for the growth of microorganisms and is used as a medium for the detection of Clostridium perfringens in foods.

#	Names of media using egg products
1	Egg albumen, soluble, dehydrated
2.	Whole egg, soluble, dehydrated (available to bacteriologists who may desire to use for studies on bacterial metabolism)
3	Egg albumen, coagulated
4	Egg yolk, coagulated
5	Whole egg, coagulated (added to another medium for growth of anaerobic organisms)
6	Egg meat medium (used to carry stock cultures of anaerobes and determination of proteolytic activity)
7	Tellurite polymyxin egg yolk (TPEY) (developed for the isolation of Staphylococcus)
8	Petrognani medium
9	Chick embryo tissue (a sterile, whole, minced chick embryo, freeze-dried for use in tissue-culture work)
10	Chick embryo extract (whole, undiluted extract of use in tissue culture procedures)
11	Petroff medium (glycerolated egg medium with beef infusion base to use with other media for isolation of turbercle bacilli)

26.4 Medical and Pharmaceutical Uses

The viable egg is essential in the production of numerous vaccinations, including those for Influenza, Pigeon pox, Epidemic typhus, Yellow fever,murme typhus,

Louping ill, Q fever, and Canine distemper. In addition to numerous pox illnesses, lymphogranuloma, and influenza vaccinations for humans, there are also vaccines for many poultry diseases, vesicular stomatitis, and infectious encephalomyelitis for horses.Other animal diseases treated with virus vaccines grown on chicken embryos include blue tongue and Rift Valley fever in sheep, and infectious bronchitis in poultry.

26.4.1 Purified Proteins

Biochemical supply companies offer a variety of pure proteins generated from eggs to researchers. ovomucoid, lysozyme chloride, avidin, ovomucoid, albumen, crystalline (egg white), phosvitin and conalbumin are among the most recently mentioned proteins. Another product for use in medicines is lysozyme.

26.4.2 Artificial Insemination

For many years, egg yolk buffered at pH 6.75 was used as the pabulum for spermatozoa in a procedure for the preservation of bull semen. Through this, vigorous spermatozoa have consistently had their ability to fertilize preserved for lengths of time exceeding 100 hours when stored at 10°C.

26.5 Fertilizer

Hatcheries and processing facilities have disposed of a large number of eggs and shells over the years. These have been ground and dried to be used as feed and fertilizer. The quality of the air and water, as well as the biological and chemical oxygen requirements of waste products, have all been taken into account in recent decades. The ammonification of soil is enhanced by albumen, according to early studies, however, it is too expensive to be used on a wide basis. The use of egg waste as fertilizer (other than as a means of waste disposal) cannot be justified in view of current costs. These surpluses might be used by the organic gardener.

26.6 Animal Feeds

26.6.1 Feeds for Domestic Animals

The majority of domestic animals' uses for eggs are related to substituting feed for other ingredients (replacement). Except in isolated cases when eggs are in surplus, only eggshell waste or eggshell meal has been economically feasible as feedstuff replacement. Both commercial egg-breaking operations and broiler or egg-type chick hatcheries produce considerable amounts of these byproducts. Eggshell waste is either obtained by mixing the liquid that typically adheres to eggshells or as the shell portion is centrifugally separated from adhering liquid egg.

26.6.2 Pet Foods

In today's supermarkets, dog and cat foods constitute a major and increasing source of eggs. Many of these foods contain poultry by-products which often contain eggs. Whole eggs are included as one of the ingredients in some speciality products for raising puppies and other animals.

26.7 Manufacturing Industrial Uses

26.7.1 Shampoos and Cosmetics

Eggs have played an important part in the history of beauty culture, especially among women. Eggs, said the manager of one large beauty shop, are wonderful for the hair. They not only cleanse die hair perfectly, but also leave it gleaming and soft with a beautiful natural sheen. Eggs frequently work wonders for skin that needs to be tightened and cleaned up. Egg oil was used in Russia during die nineteenth century in the manufacturing of Kazan soap, noted as excellent for skin treatment. Egg yolk was also included in soaps in Germany. Presently eggs are included in many commercially prepared shampoos which are selling in drug store and supermarkets.

26.7.2 Adhesives

In the past an active demand developed for technical albumen, a product recovered by centrifugally extracting the remaining albumen from crushed eggshells as they come from the breaking room. About 1 lb of technical albumen is recovered from the shells for each 30-dozen case broken out.The albumen was then dried. This technical albumen was used to fasten cork inserts in metal caps on beverage bottles. However, in recent years, dried albumen that has been prepared for human consumption has replaced technical albumen due to the relatively low prices of whites.

26.7.3 Leather

Eggs which are unfit for human consumption were packaged separately as Tanner's Egg Yolk, although the product was usually a whole egg mixture. Formerly used as a dressing for leather and furs because of the excellent emulsifying capability of yolk components, Tanner's Egg Yolk is no longer used for this purpose. Frozen and dried, Tanner's Egg Yolk was used to a limited extent in chinchilla food pellets.

26.8 Other Uses of gg

Artistic and cultural uses include Egg Decoration, eggshell Mosaics, and Tempera. Egg decorating is an ancient art that originated before the time of Christ. Egg ornamentation is still done often, although in a less expensive manner. One of the most popular art types is the traditional ornamentation so valued in the past by European royalty. Other egg art methods include scratch work, which involves painting eggshells in earth tones or using natural dyes, then scratching patterns into the painted surface. Another technique is known as wax resist, which involves applying layers of beeswax to the egg to create a design before dying it in different colours. Egg shell mosaics, which include the use of eggshell fragments for mosaic creation, are a unique art form. Some artists still use the tempera painting technique to create unusual effects. Additionally, it is also used in rubber manufacture, dye mordants in textiles, synthetic fibers, painter's ink, and in the production of photographic plates.

27

Traditional Egg Products

Bhaskar Reddy G.V.

Abstract

Traditional/indigenous egg products are very popular because of their ease of preparation and sensory attributes unique to certain egg products. Considerable progress has been made in standardization of egg product profile and mechanization of traditional egg products in India. As the demand for traditional/heritage egg products is ever growing due to rapid urbanization and industrialization and a lot of efforts need to be made to meet such increasing requirements. Processing, quality characteristics, presents consumption as well as demand scenario, underlying problems and prospects of some important traditional egg products such as mutta dosa, deemer patudi, dim kosha, egg bonda, nargisi kofta, egg roast, egg bhurji, egg keema, akuri and anda chop etc, and some popular global egg products has been discussed.

27.1 Introduction

Eggs containing nutrients that play fundamental roles beyond basic nutrition in consumers and their promotion as functional foods should be considered. Eggs rich in protein and offer a moderate calorie source (about 150 kcal/100 g), rich in vitamins and minerals and also great culinary versatility and low economic cost, which make eggs within reach to most of the population. Eggs are also relatively rich in fat-soluble compounds and can, therefore, be a nutritious inclusion in the diet for people of all ages and at different stages of life. In particular, eggs may play a particularly useful role in the diets of those at risk of low-nutrient intakes such as the elderly, pregnant women and children. Additionally, it must be mentioned that eggs can be consumed throughout the world, having no use of restrictions on religious grounds.

Traditional egg products and dishes are traditional in nature, and may have a historic precedent in a national dish, regional cuisine or local cuisine. Traditional egg products may be produced as homemade, by restaurants and small manufacturers, and by large food processing plant facilities. In India, traditional/indigenous egg products are very popular because of their ease of preparation and sensory attributes unique to certain egg products. Considerable progress has been made in standardization of egg product profile and mechanization of traditional egg products in India. Number of traditional egg products is being prepared in India to be consumed at household or at restaurant levels. India is having so many ethnic groups and each group having their own traditions, cultures, and eating habits that have developed a broad range of meat items with distinct flavor profiles. These items differ from one area to the next and from one location to the next. Traditional or indigenous egg products are those that are primarily native in origin. Although there is no written record on the methods of preparation and technological implications of traditional egg products, the procedure for preparing these items involves concepts of different contemporary processing techniques. People make decisions depending upon the availability of raw materials in different seasons of the year. The developed tastes for certain culinary items based on the availability of ingredients, and in the process, people have produced numerous specialty items typical to their regions, resulting in products that differ from region to region and place to place. The different traditional meat products of different regions of India have presented in Table 27.1 a and the brief discussion of traditional egg products has been discussed below.

Table 27.1: Regional Traditional Egg Products in India.

S.No	Egg Products	Region
1.	Egg Dosa/Mutta Dosai, Omelette Curry & Egg Kalakki	Tamil Nadu
2.	Egg Bonda	Andhra Pradesh, Karnataka
3.	Kodi Guddu Gasagasala Kura and Kodi Guddu Beerakaya Kura	Andhra Pradesh
4.	Deemer Patudi and Dim Kosha	West Bengal
5.	Nargisi Kofta	Awadh, Utter Pradesh
6.	Egg Paratha	Delhi
7.	Nadan Mutta Egg Roast, Egg Cutlet and Egg Puff	Kerala
8.	Egg Keema	Gujarat
9.	Egg Bhurji and Baida Roti	Maharastra
10.	Akuri	Western India
11.	Ros Omelet	Goa
12.	*Alu-Koni Pitika*	Assam
13.	*Anda Chop*	Orissa

27.2 Different Traditional Egg Products

27.2.1 Egg Vindaloo (Spicy Goan Egg Curry)

Egg vindaloo is a special Indian egg dish made with spices and vinegar. Preparation for egg vindaloo requires hard boiled eggs, ginger, garlic, onion, tomato, green chilies, red chilies, cumin, cloves, cinnamon, turmeric, garam masala, sugar, vinegar, lemon juice, and coconut milk. The chilies, cumin, vinegar, and lemon juice give the dish a spice and acidity that is tempered by the sweet sugar and silky coconut milk.

Cutting your eggs in half and making a puree with the garlic, ginger, tomatoes, and chilies. Next, saute the spices until they become fragrant and after a few minutes add the remaining ingredients and gently place the egg halves in the mixture. Stir the dish to combine the flavors, and allow it to simmer until the gravy thickens. Egg vindaloo serve with at any meal includes naan, steamed rice, or roti.

27.2.2 Egg Biryani

Egg biryani is a quick and easy Indian dish made by using a pressure cooking method. The resulting dish is full of deep masala flavors and spices. Egg biryani also includes lighter ingredients like fresh mint to brighten up the biriyani colour.

Preparation of egg biriyani requires eggs, chili powder, turmeric, cinnamon, cloves, garlic, onion, tomato, garam masala, mint leaves, basmati rice, and yogurt to thicken the gravy and bring out the biryani flavors. During preparation of egg biriyani, stir-fry all ingredients with the hard boiled eggs and cook them in the biryani gravy before adding the rice and cooking it directly in the gravy.

27.2.3 Egg Kurma

Egg kurma also known as egg korma and it is a delightful gravy recipe that combines savory ingredients like tomatoes and green chilies with sweeter flavors like coconut and almond paste. Mixing these ingredients and adding hard boiled eggs creates a well balanced dish with full of flavor and a variety of nutrients.

Preparation of a typical egg kurma requires eggs, green chilies, ginger-garlic paste, tomatoes, chili powder, garam masala, coriander, coconut, almond, and cashew paste. Make the egg kurma by sauteing the green chilies, sliced onions, ginger-garlic paste, and tomatoes before tossing in the rest of the spices. Then add the pastes and water until reaching desired thickness of kurma. Finally, place the hard-boiled eggs in the pan and allow to simmer together, deepening the dish's flavors. This simple dish consisting of complex flavors can be served with a side dish, rice and naan.

27.2.4 Methi Anda

Methi anda as Indian style scrambled eggs and this simple dish takes traditional scrambled eggs up a notch by spicing them with aromatic ingredients like fenugreek, garam masala, and fresh cilantro. Methi anda can be served with naan, roti, or a couple of pieces of regular toast.

Preparation of methi anda requires eggs, butter, onion, fenugreek leaves, tomato, green chilies, coriander, cumin, turmeric, chili powder, garam masala, and cilantro. Initially heat the onion, fenugreek leaves, tomato, and green chilies in a pan and add the spices before cracking the eggs and cooking them like regular scrambled eggs. Once the eggs are cooked, garnish them with garam masala and cilantro. This Indian traditional egg dish is best as a quick breakfast and some times methi anda served with an Indian condiments, such as chutney.

27.2.5 Egg Pulao

Egg pulao is an Indian egg recipe that involves cooking the rice along with the rest of the ingredients. In egg pulao preparation, basmati rice will be cooked into flavorful, spiced gravy to create a tasty richness to egg pulao. While similar to egg biryani, egg pulao requires fewer ingredients and is ready in even less time, making for a convenient and delicious meal.

For preparation egg pulao requires eggs, basmati rice, cumin, anise, ginger-garlic paste, cardamom, cloves, bay leaf, onion, fenugreek, coriander, garam masala, turmeric, red chili powder, and yogurt. Initially cook the basmati rice and boiled the eggs, then heat all spices in a large pot and saute until fragrant before adding chopped vegetables. Allow everything to simmer before adding the curd to thicken the gravy. Then, gently stir hard-boiled eggs, cooked rice, and garnish with coriander leaves

27.2.6 Shahi Egg Curry

Shahi egg curry is made up with cashew and tomato egg curry which is scrumptious when served with naan, roti, or rice. The cashew paste and tomato puree also give this egg curry a beautifully creamy consistency.

Preparation of shahi egg curry requires eggs, tomato puree, onions, garlic, ginger, bay leaf, turmeric, chili powder, garam masala, fenugreek leaves, and cashew paste. Make a paste out of the ginger and garlic, then saute it with the onions. Add the rest of the ingredients, aside from the fenugreek leaves, before covering the pan and allowing everything to simmer. Once the curry is done cooking, stir in the fenugreek leaves to complete the dish's complex flavor profile.

27.2.7 Egg Dosa

Egg dosa is a popular street food from South India and commonly referred to as mutta dosa. Dosa batter is essentially a crepe batter made from fermented rice and lentils and this batter is cooked simultaneously with scrambled eggs, chopped veggies, and spices.

For preparation of egg dosa requires, eggs, dosa batter, green chilies, chopped onion, coriander, grated carrot, salt, and oil. After preparing all ingredients, heat up a pan and ladle a scoop of dosa batter onto it. Once the batter is slightly cooked, crack an egg in the center, spread it around the dosa, and top it with green chillies and chopped onions. Once the bottom of the dosa is cooked, flip it to the other side until the egg is cooked through and the dosa is crisp. Generally egg dosa served with chutney, sauce, or any thick curry.

27.2.8 Egg Omelet

Egg omelet/Indian masala omelet is a traditional egg-based dish originating from South India. This dish is made with a combination of eggs, onions, green chili peppers, coriander, chili powder, and turmeric powder. The eggs are whisked with the other ingredients, and the mix is then cooked on both sides in a pan until the omelet becomes slightly crisp on the edges.

This spicy omelet is traditionally served for breakfast and it's recommended to eat it with pav (bread). If desired, it can be rolled in hot chapati flatbread. The dish is also often prepared & sold at street stalls in the area, and some vendors add ingredients such as tomatoes, coriander leaves, and cheese to the omelet.

Figure 27.1: Illustrations of different traditional egg products

27.2.9 Egg Bhurji

Egg bhurji is also known as Egg khagina or Anda bhurji. Egg bhurji is a traditional egg-based dish originating from India, where it's also known as Anda bhurji. It consists of spicy scrambled eggs. Although there are a few variations, the dish is usually made with a combination of eggs, oil, onions, cumin seeds, chili peppers, garam masala spices, red chili powder, turmeric, ginger, garlic, and coriander leaves. When the eggs become fully cooked and slightly dry, the dish is ready to be served, and it's often garnished with coriander. It's recommended to serve egg bhurji with pav (bread), chapati, or rice. If desired, the dish can be enriched with the addition of tomatoes and bell peppers. It can be found throughout India, from street food stalls to highway rest stops.

27.2.10 Kerala Egg Roast

Kerala egg roast also known as Nadan Mutta Roast or Mutta Roast and it is a traditional Indian dish originating from Kerala. This dish is usually made with a combination of eggs, ginger, garlic, onions, oil, mustard seeds, curry leaves, turmeric, red chili powder, lime juice, salt, and pepper. The eggs are hard boiled, peeled, and incised with a knife on the sides so that the masalas can seep into the slits. The mustard seeds are fried in oil and mixed with ginger, garlic, onions, curry leaves, turmeric, chili powder, water, lime juice, and pepper. The boiled eggs are placed into the pan and covered with the masala, then stir-fried for a few minutes. Mutta roast is often served with Indian flatbreads and basmati rice on the side.

27.2.11 Crab Omelet (Nandu Omelet)

Nandu omelet is a traditional omelet originating from India. It is usually made with a combination of eggs, crabmeat, onions, tomatoes, curry leaves, hot peppers, turmeric, oil, salt, and pepper. The crab is cleaned, steamed, and mixed with sauted onions, tomatoes, curry leaves, hot peppers, and spices. Beaten eggs are added to the mixture, and the omelet is then cooked in oil on both sides until golden brown.

27.2.12 Ros Omelet

Ros omelet is also known as Ras Omelette or Ros Omelette and it is a street food dish consisting of an omelet that is drizzled over with gravy that is based on chicken, chickpeas, or coconut milk. The omelet is typically prepared with beaten eggs, coriander, tomatoes, onions, and hot chillis. The dish is a staple of Goan cuisine and is best consumed fresh from the street carts that are dispersed along the streets of Goa, which typically come alive in the evenings. Traditionally, the dish is garnished with finely chopped onions and a splash of lime juice, while the usual accompaniment is Goan bread known as *pao*.

27.2.13 Akuri

This dish is well popular in Parsi community in Inida. This is very similar to egg bhurji, and there are two key differences: First, the eggs are intended to be runny (this is less of a hard and fast rule for bhurji), and second, powdered spices are subbed out for additional onions and whole green chillies.

27.2.14 Deemer Patudi

The cuisine of West Bengal shares a particular proclivity for slathering things in mustard. In this dish, boiled eggs get doused in a homemade paste of black or brown mustard seeds (like a less-processed version of the bright yellow stuff), covered with rice, wrapped in a banana leaf, and steamed until unbelievably fragrant. When open the banana leaf, the hot, steamy, delicious deemer patudi is ready to serve with any type of dish.

Figure 27.2: Illustration of different traditional egg products

27.2.15 Egg Bonda

Also known as egg bajji or egg pakora and egg bonda is a delicious traditional Indian recipe of spiced, deep-fried eggs. The light crisp on the outside of these boiled eggs gives them an excellent texture complemented by the subtle taste of the spiced batter used to fry the eggs. As a bite-sized food, egg bonda can be served as a snack or an appetizer before dinner.

Preparation of egg bonda requires eggs, gram flour, rice flour, caram seeds, red chili powder, turmeric, garam masala, and oil for deep frying. Initially, sliced the hard-boiled eggs, then set them aside before mix the flour, spices, and water to make a light batter. Once the batter is ready, heat up the oil to prepare it for deep frying. When the oil gets hot, dip the egg slices in batter, then fry them until golden. Then, garnish egg bonda with chopped cilantro, lemon slices, and diced red onion to add freshness to the plate.

27.3 Global Traditional Egg Products

At global level, many traditional egg products and dishes that are passed on through generations or which have been consumed for many generations. Some traditional egg products have geographical indications and traditional specialties in the USA, European Union designations per European Union schemes of geographical indications and traditional specialties. In some countries these traditional egg products are categorized under Protected designation of origin (PDO), Protected geographical indication (PGI) and Traditional specialties guaranteed (TSG). These standards serve to promote and protect names of vast traditional egg products across the globe.

Table 27.2: Global Traditional Egg Products.

S.No	Egg Products	Region
1.	Shakshuka	Tunisia, Middle East and North Africa
2.	Egg Drop Soup	China
3.	Huevos divorciados	Mexico
4.	Eggs en Cocotte	France
5.	Quiche Lorraine	Germany
6.	Tamagoyaki	Japan
7.	Kwek Kwek or Tokneneng	Philippines
8.	Scotch Eggs	Scotland
10.	Frittata	Italy
11.	Eggekage	Denmark
12.	Denver Omelet	USA
13.	Thai Omelet (Khai Yat Sai)	Thailand
14.	Rolled omelet or Gyeran mari	South Korean
15.	Omusoba	Japan
16.	Torta	Philippines
17.	Kaygana	Turkey

27.3.1 Frittata

Frittata is a traditional Italian flat and round egg-based dish that's closer to a Spanish tortilla than to a French omelette. It's made with a combination of whisked eggs and additional ingredients such as prosciutto (crudo or cotto), grated cheese, or sautéed onions. The frittata should be completely set, yet moist. The eggs should be beaten lightly so that the air bubbles don't form. The egg mixture is traditionally cooked in butter or a mix of butter and oil. When the butter or oil begin to foam, the mixture is cooked slowly over low heat, usually in a heavy-bottomed frying pan. When it has set and the top is still runny, it's placed under the grill until the top sets, but it shouldn't be browned.

27.3.2 Eggekage

Eggekage also known as Danish Omelet and it is a traditional Danish egg-based dish that resembles an omelet or a thick pancake. It consists of a mixture of lightly beaten eggs, milk, and flour, which is poured in a skillet and topped with various condiments. The dish is usually finished in the oven until the top is completely set and a light golden crust appears on the edges. Traditional toppings include slices of fried bacon, tomatoes, and chopped chives, but it can also include a variety of other ingredients. Eggekage is usually served as a hearty breakfast or dinner and is commonly complemented with rye bread or boiled potatoes.

27.3.3 Scotch Eggs

Scotch eggs are one of Britain's most popular snacks, consisting of hard-boiled eggs that are encased in sausage meat and coated in breadcrumbs. However, if desired, the eggs can be slightly runny. The combination of these ingredients is then deep-fried in hot oil until golden and crispy. Although they can be consumed either hot or cold, the eggs are traditionally consumed chilled, preferably with pickles and a fresh salad on the side. It is believed that the snack was invented in the 19th century, when people used to dip the eggs in boiling water combined with lime-powder disinfectant, in order to preserve them.

The process was called *scotching*, and the eggs were discolored in the process. Scotch eggs were invented as a way to mask the unappealing, yet perfectly edible eggs. There are people claiming that Scotch eggs were either posh food or a humble snack in the past, but regardless of the truth, their popularity started to decline rapidly during the 1990s, when they started appearing at gas stations across the country. In recent years, there has been something of a turnaround, as Scotch eggs started to appear on *Cool lists* in magazines and even in Michelin-starred restaurants.

27.3.4 Denver Omelet

Denver omelet also known as Western omelet and it is an American dish made with eggs, ham or bacon, onions, and green peppers. In order to prepare it, the mixture is poured into a frying pan, cooked, and folded to form an omelet. Although a plaque on California Street in Denver says that this dish was invented to mask the stale flavor of old eggs, it is more probable that the dish started not as an omelet, but as the Denver sandwich, which is actually a Denver omelet tucked between two pieces of toast.

Figure 27.3: Illustration of different global egg products

27.3.5 Thai Omelet (Khai Yat Sai)

Thai omelet or Khai yat sai is a traditional type of omelet originating from Thailand. The name khai yat sai means stuffed eggs. In order to prepare the dish, eggs should be lightly cooked, seasoned with oyster sauce or fish sauce, topped with ingredients such as spring onions, carrots, tomatoes, peas, ground beef or pork, then folded over before serving. Once folded over, it seems like the omelet is stuffed. It's recommended to serve khai yat sai on its own or with rice and chili peppers in fish sauce on the side.

27.3.6 Rolled Omelet (Gyeran Mari)

Gyeran mari is a type of South Korean rolled omelet that usually incorporates various additional ingredients. The dish is made with lightly beaten eggs, while the fillings typically include finely sliced or diced scallions, carrots, peppers, mushrooms, zucchini, cheese, ham, bacon, or nori seaweed. After the omelet has been rolled, it is typically served sliced. Gyeran mari is a popular breakfast dish, but it's also a common lunch box item, as well as a side dish (*banchan*) or *anju*—a small dish that's typically consumed with alcoholic beverages.

27.3.7 Omusoba

Omusoba is also known as *Omelet-Wrapped Yakisoba* and it is a Japanese dish consisting of yakisoba (stir-fried noodles, meat, and vegetables) wrapped inside a fluffy omelet. The dish is traditionally served with a tangy brown sauce similar to Worcestershire sauce, and it is especially popular as comfort food on cold or rainy days. Omusoba can be found throughout Japan in numerous family restaurants and Japanese brasseries known as *izakayas*.

27.3.8 Torta

Torta is a popular Filipino dish consisting of an omelet filled with ground meat and vegetables such as diced potatoes. Savory and extremely satisfying, the dish is usually served as a main course, preferably over rice. It is recommended to pair torta with condiments such as banana ketchup. There are numerous variations on the Filipino torta, such as *tortang talong* (eggplant omelet) and *tortang gulay* (onion, bell peppers, and garlic omelet). The reason for the unusual name of this dish is a strong Spanish influence in the Philippines. There is one other dish named torta in the country, but it is altogether different and refers to a small, muffin-like dessert cake.

27.3.9 Kaygana

Kaygana is a popular Turkish egg-based dish that can be found in numerous variations across the country. The basic version is typically prepared with just a few ingredients: eggs, flour, salt, and milk. The ingredients are mixed into a batter which is then fried in butter, and the final result is something between an omelet and a pancake. Kaygana is often enriched with ingredients such as peppers, feta cheese, tomatoes, bacon, anchovies, eggplants, green onions, mushrooms, and even honey for the sweet versions. It is recommended to serve the dish warm, ideally with bread and a beverage of choice.

27.4 Conclusion

Traditional egg products are part of the identity of a local culture and have unique sensory attributes to satisfy the gastronomic desires of the vast majority of a local populace in the India. The egg products consumption is increasing exponentially in different regions of India. Traditional egg products have immense potential to generate income and employment for the local people by organized production and processing of these products. At present there are many products that are not scientifically standardized and for which there is no comprehensive information available for their commercial production. The available methods of manufacture of traditional egg products in batches are time consuming and there is limited shelf life also. Thorough knowledge of their formulations, process optimization, commercial production, advanced packaging, and cold chain for distribution could definitely put many potential traditional egg products on the world map. Retort pouch processing can be applied to some of the traditional egg products from the India such as egg curries, egg pickles, egg biryanis, egg keema, egg bhurji, egg pulao and egg bonda etc., exploit their market potential. Certain products, such as egg omelet, boiled egg and egg dosa to their nutritional-rich form can be marketed as healthy foods. With better know-how and technology, the traditional egg products from the India could become more popular and more widely available to global market.

28

Shelf Stable Egg Products

Bhaskar Reddy G.V.

Abstract

Several shelf stable and precooked egg-based products are currently available in markets around the world. These products include egg pizza, scrambled eggs, omelets, French toast, breakfast sandwiches, cubes, crepes, quiche and also different dried egg powders. These products are mainly sold in frozen form and dried form to fast-food outlet chains in the food service industry. However, non-dried shelf-stable egg- based products that can be stored at room temperature are not yet available. At present, the development of shelf-stable egg-based products using a conventional retort method for sterilization has not been successful due to the green-gray discoloration of the final egg product that affects its appearance, the development of off-flavors, and the syneresis or exudation of water after heat treatment. But combining high temperature with other alternative processes like high pressure techniques and usage of ozone etc are requires for manufacturing of shelf-stable egg-based products.

28.1 Introduction

The majority of eggs produced in the world are sold in-shell for consumers to utilize in the home in traditional ways or for baking. Traditionally, surplus eggs not required for the shell egg market (normally the smaller sizes) and downgraded eggs were turned into egg products. Today, farmers have specific contracts to produce eggs for further processing, and any egg that is damaged cannot be used for this purpose. There are essentially three types of egg products, liquid, dried, and frozen, with small markets for peeled, hard-boiled, and hard-boiled pickled eggs. The main products are used widely in the bakery and confectionery trades along with other industries. Many uses to which processed whole eggs can be put and also shows the way in which whole eggs can be used to produce value-added products for the food industry without going through the process of pasteurization with or without subsequent drying has shown in Figure 28.1.

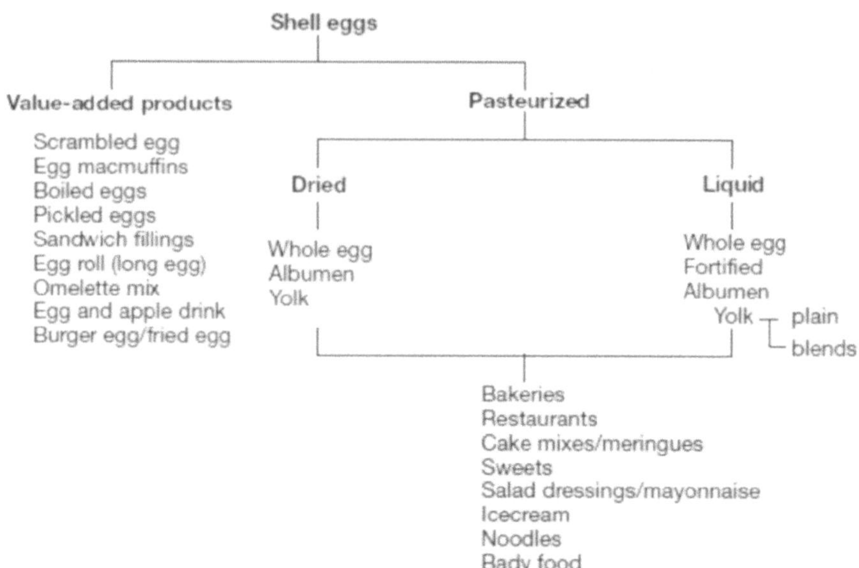

Figure 28.1: Different types of eggs in the food industry

Eggs have been considered to be highly nutritious containing high levels of vitamins and minerals. The physical appearance of an egg makes the first impression upon the consumer. If the product does not meet perceived expectations, consumer confidence diminishes. The structural quality of the shell egg is important to the processor because eggs that are structurally sound will arrive to the consumer in the best condition. Furthermore, high interior quality is of importance to egg products manufacturers because it allows for better separation of components without cross over contamination, especially when produce albumen products. Freshness is a major contribution to the egg quality. The internal quality of eggs begins to deteriorate after they have been laid due to loss of moisture and carbon dioxide via the eggshell pores.

The shelf-stable egg products are formulated by combining various edible polyhydric alcohols and either dried egg whites, dried egg yolks or dried whole egg solids. Starch hydrolysate solids and the remaining dried ingredients are next dissolved in water and the solution is then combined with the first mixture and heat set by boiling or frying to form textures similar to that of the natural egg. The product is then allowed to cool to room temperature and is then packaged. The shelf-stable egg products which are packaged in non-hermetic pouches have an appearance, texture and colour, not unlike conventional heat-set egg products. When removed from the pouch, the moist and soft product is warmed just prior to consumption.

28.2 Basic Principles Involves During Production of Shelf Stable Egg Products

Eggs serve important roles in many food products because of their diverse functional properties. The structure and composition of eggs in relation to their functional properties are considered for the production of shelf stable egg products. The three functional properties of eggs that are of importance with respect to food preparation are coagulation, emulsification, and foaming. As with other high-protein foods, the denaturation and coagulation of egg proteins are important to the functional performance of eggs in a food system. Denaturation is a rearrangement of the orderly alignment of the molecules in a native protein. Peptide chains of a protein usually are coiled into a globular compact shape. This arrangement is stabilized with weak bonds between amino acid residues along the peptide chains. During denaturation, some of the forces that hold the chains in coiled and folded positions are disrupted, allowing the peptide chains to uncoil and unfold. New bonds are formed to give a new arrangement to the protein molecules.

28.2.1 Coagulation

Coagulation is ą term used to describe that entire process that results in a loss of solubility or a change from a fluid to a more solid state. The term gelation, meaning the formation of a gel, also is used to designate the loss of fluidity of egg white and yolk. The coagulation of eggs is responsible for the thickening effect that eggs have in products such as custards. Egg white begins to thicken as the temperature reaches 62°C, and at 65°C it will not flow. At 70°C the mass is fairly firm. Yolk coagulates at a somewhat higher temperature than white. It begins to thicken at 65°C and loses its fluidity at about 70°C. Coagulation does not occur instantaneously but gradually over a period of time.The reaction proceeds more rapidly as the temperature of heating is increased. Since heat is absorbed during coagulation of egg proteins, the reaction is endothermic. The firmness of an egg white coagulum or whole egg coagulum is dependent on the time and temperature of heating. Albumen coagulum increased in toughness as temperature of heating increased from 77 to 90°C and as time of heating increased from 7 to 60 min, whereas whole egg coagulum toughened as much at 80°C as at higher temperatures.

28.2.2 Emulsification

Egg yolk, which is itself an emulsion, is a good emulsifying agent for fats or oils and water. It performs this function in many foods, including shortened cakes, salad dressing, and mayonnaise. The emulsifying qualities of eggs often are studied in mayonnaise because it is a relatively simple system. In mayonnaise, particles of yolk interact at the surface of oil droplets to form a layer that protects the emulsion by inhibiting coalescence of the oil droplets. Participants in the layer include both low- and high-density lipoproteins.

28.2.3 Foaming

The foaming of egg whites is important in many foods, because it makes the products light in texture and contributes leavening. Egg white foam is a colloidal suspension consisting of bubbles surrounded by albumen that has undergone some denaturation at the liquid-air interfaces. This denaturation, which is caused by the drying and stretching during beating, makes some of the albumen insoluble, thus stiffening and stabilizing the foam.

The globulins exhibited greater foaming ability than did ovomucin, lysozyme, conalbumin, ovomucoid, and ovalbumin. As foams were formed, the number of sulfhydryl groups decreased, reflecting sulfhydryl-disulfide interchanges during foaming. Volumes of cakes ranging from largest to smallest were produced from globulins, ovalbumin, control (a mixture of all proteins), conalbumin, lysozyme, ovomucoid, and ovomucin. Lack of aeration or instability of foams was responsible for low volumes with the last four. When combinations of proteins were studied, it was shown that the inclusion of ovomucin increased viscosity and contributed to foam formation but depressed cake volume. The addition of lysozyme with the ovomucin decreased foaming but increased cake volume, suggesting interaction between ovomucin and lysozyme. During denaturation the protein molecules unfold, and their polypeptide chains lie with their long axes parallel to the surface. Overheating incorporates too much air, resulting in extensive denaturation of ovomucin so that the protein films become thin and less elastic. Elasticity is needed, especially in foams that are to be baked, so that the incorporated air can expand without breaking the cell walls before the ovalbumin is coagulated by the heat of the oven.

28.3 Different Shelf Stable Egg Products

28.3.1 Dried Egg Products/Powders

Dried egg powders are made in the same way as powdered milk is formed. The finished product is a fine powder, which on reconstitution gives a product similar to fresh whipped eggs. Different types of end products can be made by drying different components of eggs together or separate like egg albumin (egg white) powder, egg yolk powder, mix or whole (white and yolk together) egg powder, and eggshell powder. To dry egg whites and yolks, techniques like pan-drying, foam-drying, freeze-drying, oven-drying, and spray-drying are commonly used. The spray-drying method is most commonly used for preparation of egg white and yolk powder

28.3.1.1 Pan Drying

Pan-drying is a noncommercial method used to dry eggs at home. There are two ways to dehydrate eggs; one by using raw eggs directly, also known as wet-dying, and second by using precooked eggs, also known as a dry method. The choice of method depends on various factors like convenience, product characteristics, safety and processing facilities. After procurement and the removal of shells, liquid eggs are homogenized and dehydrated followed by storage in airtight containers. In precooked methods, egg whites and yolks after homogenization are transferred to a nonstick pan for cooking over medium flame until the material becomes solid and no visible water remains. The large clumps are broken to reduce the drying time. After cooking, scrambled eggs are spread on dehydrator trays. Set the thermostat and timer of dehydrator to 65°C for 4 h for complete drying. After dehydration, dehydrated eggs are converted into a powder using blender followed by storage in airtight containers. In the case of the wet method, whipped and homogenized eggs are transferred on dehydrator trays. The thermostat of dehydrator is set on 65°C for 10–12 h. After grinding, the powdered eggs can be kept for up to a year. As powdered eggs are a nonperishable commodity, their stay time is about 5–10 years depending on the type of environment being provided if powdered eggs are being stored in airtight containers or placed where there is no oxygen or if placed in a cold room/area these can even remain longer. After dehydration, the raw eggs are orange and fine in texture. On the other hand, precooked eggs are yellow and a bit coarser. After rehydration and cooking, the raw eggs are closer to fresh eggs and the precooked eggs are grainy. Dehydrated raw eggs are good for egg products like omelets and dehydrate precooked eggs are as an ingredient like in baked products.

28.3.1.2 Foam Drying

Premium quality drying products can be prepared from foam-mat drying. In foam-mat drying, the product is converted into foam before drying. This gives an advantage of increasing the total surface area, thus improving the final drying rate. Another advantage of this method is that the drying temperature is lower than that of conventional drying methods and this helps to reduce the loss of flavor and volatile components. After procurement of raw material, eggs are separated as albumen liquid and whole egg liquid followed by homogenization and addition of stabilizers during the whipping process for the stability of foam at ambient temperature. On whipping, the proteins of egg white denature at the interface and interact with one another to form stable, visco-elastic interfacial film. Then foam is spread on a sheet or mat and dried by means of hot air at atmospheric pressure in cabinet dryer. Another way to dehydrate the egg is foam mat freeze drying in which material is passed through freezing process below −40°C. Subsequently, drying is carried out in freeze dryer for nearly 6 h.

28.3.1.3 Oven Drying

After procurement of the eggs, these are candled to confirm freshness. Good-quality eggs are cleaned and sanitized by dusting, washing with water (temperature of water should be at least 10°C warmer than the internal temperature of egg), and food-grade sanitizer. Eggs are deshelled and separated as white, yolk, and whole egg liquid, later homogenized and treated with hydrogen peroxide solution to destroy viable *Salmonella* microorganism and to prevent browning of the products, followed by oven-drying at 44°C for 4 h. After cooling and grinding of dried egg flakes, egg powder is stored at optimal condition. Results showed that functional properties like emulsifying capacities and stabilities, foaming capacities and stabilities, water and oil absorption capacities, solubility indexes and coagulation temperatures don't affect impact negatively. Whole egg powder exhibits better functional properties like foaming and oil absorption capacities. Furthermore, egg yolk powder has shown good scores in emulsifying capacities, stabilities, and coagulation temperature. Likewise, egg whites perform best in foaming stabilities, water absorption capacities, and solubility indexes. The nutritional composition compared to fresh eggs has an insignificant impact on drying temperature (44°C) on the nutritional value of the dried egg. The moisture content of these powdered materials is low enough to extend the shelf life during storage at low humidity.

28.3.1.4 Freeze Drying

Freeze-drying is a process of drying a product through freezing primarily and secondary drying. The sample is exposed to low temperatures sufficient to become solid by forming ice crystals followed by ice sublimation, driven by low pressures and temperatures, and secondary drying during which the residual water is removed by desorption to an optimal level for stability and get the dried product. In conventional freeze-drying, eggs are frozen until the central temperature reaches up to −40°C, after homogenization, followed by freeze-dried drying process for about 12 h, final moisture content of 3.5%. Freeze-drying can be combined with microwave vacuum-drying to reduce drying time. Frozen egg samples are dried for about 8 h to a moisture content of about 25%, using the same conditions as for freeze-drying. Then the samples are transferred into the microwave vacuum-drying oven, with the drying parameters set as follows: vacuum pressure 80 Pa, frequency of microwave 2450 MHz, microwave power 600W, drying pressure 5 kPa, drying time about 30 min, until a final moisture content of 3.5%.

Table 28.1: Characteristics of egg powder produced through raw dehydration and pre-cooked dehydration.

S.No	Characteristics	Methods	
		Raw Egg Dehydration	**Pre-cooked Dehydration**
1.	Powder texture	Orange, fine powder	Yellowish, powder the size of couscous
2.	Rehydration ability	Soak water completely	Don't soak water completely
3.	Cooking behavior	Stuck a bit to pan	Don't stick to pan
4.	Texture after cooking	Like fresh eggs	Grainy
5.	Taste	Like fresh eggs	Not appetizing

Source: Adapted from Erich, 2012. How to make powdered eggs. Available from: http://tacticalintelligence.net/blog/how-to-make-powdered-eggs.

28.3.1.5 Spray-drying

Spray-drying is a technique used to preserve liquid foods, by converting into powder form. In this technology feed material is sprayed into hot dry air, having temperatures ranging 100–300°C. In this case, the liquid portion of the feed material is evaporated resulting in a final product that is dried powder either in granules or agglomerates form. The structure of the final product depends on the physical and chemical properties of material, design of the dryer, and operating conditions. In spray-drying, liquid eggs (droplet 10–200 µm) with high pressure (130–200 bar or 2000–3000 psi) injected into hot air (160–194°C) to become powder in approximately 12 s. The spray-drying process yields food powders that are then encapsulated and for this purpose different encapsulated agents are used that enhance the bulk and flow ability of food powders. Glucose needs to be removed from egg whites to avoid the issue of caramelization and Maillard reaction and ensure a good stabilized product. For this purpose, various methods can be used, like including natural fermentation at 24–29°C for 6–7 days, glucose oxidase enzyme fermentation, and control bacterial or yeast fermentation by *Streptococcus* and *Lactobacillus* at 30–35°C in 24 h (unpasteurized product) or by *Sacharomyces*, respectively. Whipping aids like sodium-lauryl-sulfate (<0.1% w/w) can be added in the liquid before spraying to facilitate the whipping process. After drying hot room treatment is given to egg white powder at various temperatures for specific times like 54.4°C for 7 days, 60°C for 8 days, or 64°C for 14 days. To dry yolks and whole eggs, the material is pasteurized and desugared before drying to increase the shelf life at room temperature. To increase whipping capabilities and product stability, glucose-free corn syrup and sucrose can be added before drying. Drying targets for egg white liquid to powder is 8:1, for whole egg liquid to powder is 4:1, and for yolk liquid to powder is almost 2:1.

28.3.1.6 Advantages of Dried Egg Products

They can be handled and stored with ease and at low cost. They are ready to use immediately with no thawing.

1. They are easy to handle in a hygienic way.
2. They are easy to remove from the container without scraping.
3. No bacterial growth can occur in powder at room temperature, provided it is kept dry.
4. There is good uniformity. There is precise control over the amount of water used in formulation.
5. There is no loss when used because the dried egg is usually added directly to the batch.
6. No special transfer or storage equipment is needed.
7. Because the moisture content is reduced from around 74% to 2–4% by weight, there is a reduction in weight and volume and a concentration of food value.

A possible disadvantage of powder compared with liquid egg is a loss of 'fresh flavor' and a loss of certain functional properties such as aerating power, unless treated with non reducing sugars prior to heating. This, however, limits its use because of the sweetened nature of the powder.

28.3.1.7 Quality of Dried Egg Products

The quality of dried egg or dried egg products can be affected by a number of factors. Important are the quality of the eggs broken out to make the original product, handling methods, sanitation practices, conditions during processing, pasteurization procedures, drying, and conditions under which the products are held in storage. Glucose (0.3–0.5%) is removed from the liquid egg white prior to drying by fermentation using a yeast or bacterial culture or by oxidation to gluconic acid using a glucose oxidase–catalase enzyme system. With glucose removed, dried egg whites are completely stable. If glucose is not removed, the product would be unstable because reducing groups from the glucose would combine with amino acids in the protein, leading to a condensation reaction that would be followed by browning and the development of insoluble proteins. There could also be the development of off-odors and the loss of some functional properties during storage. The physical properties important in relation to dried egg products are bulk density, dispersibility, solubility, and reconstituted viscosity.

28.3.1.8 Storage of Dried Egg Products

One of the practical advantages of dried egg products is their ease of storage. Most are relatively stable when stored at room temperature. Dried egg whites can be held under almost any storage condition for an indefinite period of time. Dried products containing whole egg and yolk should be under refrigeration if held for long periods of time. Some of these are relatively stable at room temperature, particularly those where the natural glucose has been removed prior to drying. Generally dried whole egg solids should be kept cool at less than 50°F (10°C) to maintain quality. Once containers of egg solids have been opened, reseal tightly to prevent contamination and absorption of moisture. And once dried eggs are reconstituted, use immediately or store refrigerated for not more than four days. If dried eggs are combined with dry ingredients and held for storage, they should be sealed tightly in a closed container and stored in the refrigerator at 32° to 50°F (0° to 10°C). Plain unstabilized whole egg solids have a shelf life of about one month at room temperature and about a year at refrigerated temperatures. Store dried eggs as any other dried, powdered food in a cool and dark place.

28.3.2 Hard-Boiled Eggs

Although liquid eggs and dried eggs still account for the majority of eggs used in the food industry, there are other products, which have evolved during the last 10–15 years, that are commercially available are hard boiled/cooked eggs. The mass production of hard-boiled eggs has developed in recent years. Normally, white eggs are used in the process, rather than brown, because the shells can be more easily removed from white eggs than from brown, resulting in less damage to the finished product. Again, only good-quality eggs are used in this process so that the final product is also of good quality. Normally, in small eggs should be used (under 53 g) for boiling as there is little demand for these sizes of eggs in the shell egg market.

The eggs are gradually boiled as they move around a trough arrangement containing high-temperature water, with the objective of them being hard-boiled by the time they have completed a single circuit of the trough. The eggs are then removed, deshelled, and cooled. Once hard-boiled, the eggs can be used for a number of products. They can be pickled in jars in plain vinegar or with spices. They can be used in pies or salads and a major use in this form is in airline meals. Eggs damaged during the boiling or shelling processes would not necessarily be discarded. They can be diverted to the manufacture of sandwich fillings, such as egg mayonnaise, which can be packed into 5 kg tubes for distribution to sandwich manufacturers and the catering trade.

The length of time required for coagulation of eggs cooked in the shell depends on the heating temperature. When eggs were boiled at unusually low temperatures (85°C for 30 min) the yolk was not firm enough for slicing and the white did not coagulate firmly after 90 min at 72°C. The temperature at which eggs are boiled/cooked in the shell is varied by starting the eggs in cold water and bringing to a boil, or by starting them in boiling water and turning off the heat. In either method, the proportion of water to eggs and the size of the pan must be controlled for uniform results. Both methods produced satisfactory products, but the boiling water start was preferred because there were fewer cracked eggs and the eggs were easier to peel. In addition, drilling or piercing the shell before cooking increased cracking during cooking.

28.3.3 Poached and Fried Eggs

Poached eggs are prepared by dropping broken out eggs into boiling water. If the white is thin, it may be difficult to obtain a satisfactory product with any method. The addition of vinegar or salt to the cooking water hastens coagulation, and hence may improve the shape of poached eggs when they have thin whites. The shape of a fried egg also depends on the viscosity of the white and on the temperature at which it is cooked. If the temperature of the cooking utensil is too low (115°C), the egg will spread excessively, and if it is too high, (146°C), the egg will be overcooked. Generally the temperatures of 126 to 137°C are satisfactory for production of poached and fried eggs.

28.3.4 Scrambled Eggs

During preparation of scrambled egg, first heat a liquid egg product to a temperature less than a cooking temperature for the egg, but a temperature which is sufficiently high to prevent heat shock of the liquid egg product and prevent non-uniformity in the scrambled eggs produced by the process to provide an initially heated liquid heated egg product. Further heat the initially heated liquid egg product to provide a homogeneously heated egg, the further heating to a temperature effective for cooking the egg and effective for causing the heated liquid whole egg product to coagulate when the liquid egg product is held in a holding tube, then transport the homogeneously heated liquid egg product to a holding tube and kept for an amount of time effective for forming a fully coagulated egg then remove the fully coagulated egg from the holding tube to form a scrambled egg product. Overheating of eggs results in a firm, rubbery mass if little or no liquid is added, or in a curdlike mass that may separate during cooking if much liquid is used.

28.3.5 Pickled Eggs

A simple, cost-effective and efficient technology are available for development of pickling of eggs up to 4 months of storage and marketing at ambient temperature in ready-to-eat form. Pickled eggs are typically hard boiled eggs that are cured in vinegar or brine. As with many foods, this was originally a way to preserve the food so that it could be stored months later. Pickled eggs have since become a favorite among many as a snack items popular in many parts of the world. After the eggs are hard boiled, the shell is removed and they are submerged in a solution of vinegar, salt, spices, and other seasonings. Pickle formulations vary from the traditional brine solution for pickles, to other solutions, which can impart a sweet or spicy taste. The final taste is largely determined by the pickling solution. The eggs are left in this solution from one day to several months. Prolonged exposure to the pickling solution may result in a rubbery texture. A common practice is to puncture the egg with a toothpick to allow the pickling solution to penetrate to the egg's interior, but this is not advised as it can introduce bacteria into the finished product.

28.3.6 Albumen Rings

Albumen rings are egg snack food, prepared by cooking blended egg albumen in ring moulds, battering and breading the coagulated albumen prior to deep fat frying. It can be popularized as egg snacks at growing fast food outlets. It contains 11.5% protein and merely 3.2% fat and also shelf life of 12-18 days at refrigeration (4 degrees).

28.3.7 Egg Roll

It is a nutritious, tasty and convenient egg product suitable for meals or as snack foods. This product offers a potential market for growing fast food outlets. The egg roll filled with 80% scrambled egg and 20% chicken meat mixture (shallow pan-fried) was rated best in flavour, texture and overall acceptability. The egg roll had a refrigerated shelf-life of 8 days in the vacuum and 6 days in the aerobic pack. Shelf life of egg roll in refrigerated storage at (5 degrees) for 8 days in vacuum and 6 days in aerobic packaging.

28.3.8 Egg Crepe

Egg crepe is a thin, fat, circular product and may be filled with meat or vegetables and rolled or folded. It is an egg-rich product and can be popularized as a convenient egg item at growing fast food outlets and at homes. Crepes has a shelf-life of 22 days in vacuum and 20 days in aerobic packaging at refrigeration (4 ± 1°C) and for 60 days at freezing (-18 ± 1°C) temperature in both vacuum and aerobic packaging.

28.3.9 Egg Waffles

Egg Waffle is a nutritious, light, crispy and versatile snack food for the breakfast. This product offers a potential market at growing fast food outlets. Egg Waffles prepared from 65% liquid whole egg with 10% wheat flour and 5% granulated wheat are most acceptable and has an ambient shelf-life of 4 days in a vacuum and 3 days in air packs, while at refrigeration temperature, it can keep well for 10 days in vacuum and 6 days in air packs with satisfactory microbiological quality.

28.3.10 Egg Cube

A batch of 100 eggs (50-55 g each), kept under room temperature (28±2°C) were broken over a sieve to separate albumen and yolk. The liquid egg albumen was mixed with a wire balloon whisk in Hobart mixer for 4-5 min to get the maximum fully blown foam. Later egg yolk was added and mixed for 1-2 min. Malic acid and citric acids were added at the beginning of beating, when the egg yolk starts becoming frothy. The binder mix containing garlic powder, wheat flour, corn starch, maltodextrin and soya protein was added slowly to homogenize the liquid egg while mixing continuously to obtain a batter with homogenous consistency. The mixing was carried out for 3-4 min till a uniform smooth batter was obtained. The resultant batter was transferred to rectangular stainless-steel lined with polypropylene sheet and the product was steamed about 35 min (internal temperature 86±2°C) to obtain solidified product followed by cooling at ambient temperature (28±2°C) for 30-45 min. The loaf was cut into cubes by using a cutting mould of 1x1x1 cm. size. The resultant product was then dried in a cross-flow dryer by spreading in stainless steel trays at the rate of 0.75 kg m^{-2}.The drying was carried out at 82±3°C for 3 h to obtain dried cubes called egg cubes. The dried EC were allowed to cool. After cooling, the product was then packed in metalized polyester (polyester 10-11 micron/aluminium foil 9-12 micron/polythene 100 gauges) bags of 50 g capacity each and stored at ambient temperature (28±2°C). Generally the shelf life egg cubes at ambient temperature is about 4 months.

Figure 28.2: Illustration of different shelf stable egg products

28.4 Conclusion

Producing a thermally sterilized egg-based product with increased shelf life without losing the sensory and nutritional properties of the freshly prepared product is challenging. Until recently, all commercial shelf-stable egg-based products were sterilized using conventional thermal processing; however, this heat treatment was problematic as quality attributes of the product were negatively affected. Combining high temperature with other alternative processes like high pressure techniques, usage of ozone etc are requires manufacturing shelf-stable egg-based products. Further, lot of attention is required to product improvement in terms of overall quality through modification of ingredients, formulation, processing and packaging, technology viable to commercial application to produce egg-based shelf stable products with increased shelf life and appealing quality to the consumer.

Index

R

S

T

U

V

W

X

Y